Lecture Notes in Computer Science 1391

Edited by G. Goos, J. Hartmanis and J. van Leeuwen

Springer
Berlin
Heidelberg
New York
Barcelona
Budapest
Hong Kong
London
Milan
Paris
Santa Clara
Singapore
Tokyo

Wolfgang Banzhaf   Riccardo Poli
Marc Schoenauer   Terence C. Fogarty (Eds.)

Genetic Programming

First European Workshop, EuroGP'98
Paris, France, April 14-15, 1998
Proceedings

Springer

Series Editors

Gerhard Goos, Karlsruhe University, Germany
Juris Hartmanis, Cornell University, NY, USA
Jan van Leeuwen, Utrecht University, The Netherlands

Volume Editors

Wolfgang Banzhaf
University of Dortmund, Department of Computer Science
Joseph-von-Fraunhofer-Straße 20, D-44227 Dortmund, Germany
E-mail: banzhaf@cs.uni-dortmund.de

Riccardo Poli
University of Birmingham, School of Computer Science
Edgbaston, Birmingham B15 2TT, UK
E-mail: r.poli@cs.bham.ac.uk

Marc Schoenauer
École Polytechnique, Centre de MathématiquesAppliquées
F-91128 Palaiseau Cedex, France
E-mail: marc.schoenauer@polytechnique.fr

Terence C. Fogarty
Napier University, Department of Computer Studies
219 Colinton Road, Edinburgh EH14 1DJ, UK
E-mail: tcf@dcs.napier.ac.uk

Cataloging-in-Publication data applied for

Die Deutsche Bibliothek - CIP-Einheitsaufnahme

Genetic programming : first European workshop ; proceedings /
EuroGP '98, Paris, France, April 14 - 15, 1998. Wolfgang Banzhaf ...
(ed.). - Berlin ; Heidelberg ; New York ; Barcelona ; Budapest ; Hong
Kong ; London ; Milan ; Paris ; Santa Clara ; Singapore ; Tokyo :
Springer, 1998
  (Lecture notes in computer science ; Vol. 1391)
  ISBN 3-540-64360-5

CR Subject Classification (1991): D.1, F.1, I.5, J.3

ISSN 0302-9743
ISBN 3-540-64360-5 Springer-Verlag Berlin Heidelberg New York

© Springer-Verlag Berlin Heidelberg 1998
Printed in Germany

Typesetting: Camera-ready by author
SPIN 10636918   06/3142 – 5 4 3 2 1 0   Printed on acid-free paper

Preface

Evolutionary Computation (EC) holds great promise for computer science today. After an early start in the 1950s, it was pursued by a handful of scientists until it took off as a rapidly growing field in the 1980s.

Evolutionary computational approaches solve problems in various application domains which can be cast as abstract search spaces, defined by those problems, to be traversed by search processes. The common feature of all branches of EC is that they are path-oriented search methods, i.e., a variety of candidate solutions is visited and stored as starting points for further exploration. Exploration takes place mostly by stochastic means, although deterministic moves are also possible.

One of the forefathers of the field, Hans Bremermann, wrote in 1962: "The experiences of various groups who work on problem solving, theorem proving, and pattern recognition all seem to point in the same direction: These problems are tough. There does not seem to be a royal road or a simple method which at one stroke will solve all our problems. ... Problems involving vast numbers of possibilities will not be solved by sheer data processing quantity. We must look for quality, for refinements, for tricks, for every ingenuity that we can think of. Computers faster than those of today will be of great help. We will need them. However, when we are concerned with problems in principle, present day computers are about as fast as they will ever be."

Nature provides a rich source of tricks and refinements. It is in this spirit that EC came into being, deriving from Darwinian evolution in biology. Biological notions such as population, mutation, recombination, and selection have been transferred and put to use in more abstract computational contexts.

Genetic Programming (GP), the youngest branch of EC, has grown rapidly since the publication of a book on the subject by John Koza in 1992. More than 800 papers have been published over the last few years. GP can be considered one of the few methods of automatic programming since the structures of the population being evolved are computer programs.

GP has already been applied successfully to a large number of difficult problems like automatic design, pattern recognition, robotic control, synthesis of neural networks, symbolic regression, music and picture generation, and many others.

This volume contains the proceedings of EuroGP'98, the First European Workshop on Genetic Programming held in Paris, France, on 14-15 April, 1998. EuroGP'98 was the first event entirely devoted to genetic programming to be held in Europe. The aims were to give European and non-European researchers in the area of GP, as well as people from industry, an opportunity to present their latest research and discuss current developments and applications. The workshop was sponsored by EvoNet, the Network of Excellence in Evolutionary Computing, as one of the activities of EvoGP, the EvoNet working group on genetic programming.

Eighteen papers were accepted for publication in this volume and for presentation at the workshop (twelve for oral presentation, six as posters). Many of these are by internationally recognized researchers in GP and EC; all are of a high quality. This has been ensured by an international program committee including not only the main GP experts in Europe but also several leading GP researchers from around the world.

We are extremely grateful to them all for their thorough work which has allowed us to provide three independent anonymous reviews for each paper submitted despite the very short time available. With such a high-quality international program committee, with the invited speech given by John Koza, the founder of GP, and with authors coming from thirteen different countries (Belgium, Brazil, Bulgaria, France, Germany, Ireland, Italy, Japan, Spain, Sweden, The Netherlands, UK, USA), ten of them European, we believe that the workshop and these proceedings represent a cross section of the best genetic programming research in Europe and in the rest of the world.

April 1998   Wolfgang Banzhaf, Riccardo Poli,
                                        Marc Schoenauer, and Terence C. Fogarty

Organization

EuroGP'98 is organized by EvoGP, the EvoNet Working Group on Genetic Programming

Organizing Committee

Program Co-chair: Wolfgang Banzhaf (University of Dortmund, Germany)
Program Co-chair: Riccardo Poli (University of Birmingham, UK)
Local Chair: Marc Schoenauer (Ecole Polytechnique, France)
Publication Chair: Terence C. Fogarty (Napier University, UK)

Program Committee

Peter Angeline, Natural Selection, New York, USA
Wolfgang Banzhaf, University of Dortmund, Germany
Tobias Blickle, Saarbruecken, Germany
Marco Dorigo, Free University of Brussels, Belgium
Gusz Eiben, University of Leiden, The Netherlands
Terence C. Fogarty, Napier University, UK
Frederic Gruau, Center for Information Science, The Netherlands
Hitoshi Iba, Electrotechnical Laboratory, Japan
William B. Langdon, The University of Birmingham, UK
Jean-Arcady Meyer, Ecole Normale Superieure, France
Peter Nordin, DaCapo, Sweden
Una-May O'Reilly, Massachusetts Institute of Technology, USA
Riccardo Poli, University of Birmingham, UK
Marc Schoenauer, Ecole Polytechnique, France
Michele Sebag, Ecole Polytechnique, France
Andrea Tettamanzi, Genetica, Italy
Marco Tomassini, University of Lausanne, Switzerland
Hans-Michael Voigt, Center for Applied Computer Science, Berlin, Germany
Byoung-Tak Zhang, Seoul National University, Korea

Sponsoring Institutions

Ecole Polytechnique, France.
EvoNet: the Network of Excellence in Evolutionary Computing.

Table of Contents

A Review of Theoretical and Experimental Results on Schemata in Genetic Programming

Riccardo Poli and W. B. Langdon

School of Computer Science
The University of Birmingham
Birmingham B15 2TT, UK
E-mail: {R.Poli,W.B.Langdon}@cs.bham.ac.uk

Abstract. Schemata and the schema theorem, although criticised, are often used to explain why genetic algorithms (GAs) work. A considerable research effort has been produced recently to extend the GA schema theory to Genetic Programming (GP). In this paper we review the main results available to date in the theory of schemata for GP and some recent experimental work on schemata.

1 Introduction

Genetic Programming (GP) has been applied successfully to a large number of difficult problems [8, 7, 2]. However a relatively small number of theoretical results are available to try and explain why and how it works.

Since John Holland's seminal work in the mid seventies and his well known schema theorem (see [6] and [3]), schemata are often used to explain why GAs work (although their usefulness has been recently criticised, e.g. in [4] and [1]). In particular it is believed that GAs solve problems by hierarchically composing relatively fit, short schemata to form complete solutions (building block hypothesis). So the obvious way of creating a theory for GP is to define a concept of schema for parse trees and to extend Holland's schema theorem.

One of the difficulties in obtaining theoretical results using the idea of schema is that the definition of schema for GP is much less straightforward than for GAs and a few alternative definitions have been proposed in the literature. All of them define schemata as composed of one or multiple trees or fragments of trees. In some definitions [8, 10, 11, 18, 19] schema components are *non-rooted* and, therefore, a schema can be present multiple times within the same program. This, together with the variability of the size and shape of the programs matching the same schema, leads to some complications in the computation of the schema-disruption probabilities necessary to formulate schema theorems for GP. In more recent definitions [14, 17] schemata are represented by *rooted* trees or tree fragments. These definitions make schema theorem calculations easier. In particular, in our work [14] we have proposed a new simpler definition of schema for GP which is much closer to the original concept of schema in GAs. This concept of schema suggested a simpler form of crossover for GP, called one-point crossover, which allowed us to derive a simple and natural schema theorem for GP with one-point crossover. We will critically review the main results obtained to date in the theory of schemata for GP in Sect. 3 after briefly recalling and reformulating Holland's schema theory for binary GAs in Sect. 2.

Although theoretical results on schemata are very important, it is well known that schema theorems only model the disruptive effects of crossover and represent only short-term predictions. At the time of writing this paper, only one empirical study on GP schemata had been carried out [12]. This analysed the effects of standard crossover, one-point crossover and selection only on the propagation of schemata in real, although small, populations. We describe the main results of this study in Sect. 4 and we draw some conclusions in Sect. 5.

2 Background

As highlighted in [16], a schema is a subspace of the space of possible solutions. Usually schemata are written in some language using a concise notation rather than as ordinary sets, which would require listing all the solutions they contain: an infeasible task even for relatively small search spaces.

In the context of binary representations, a schema (or similarity template) is a string of symbols taken from the alphabet $\{0,1,\#\}$. The character $\#$ is interpreted as a "don't care" symbol, so that a schema can represent several bit strings. For example the schema $\#10\#1$ represents four strings: 01001, 01011, 11001 and 11011.[1] The number of non-$\#$ symbols is called the *order* $\mathcal{O}(H)$ of a schema H. The distance between the furthest two non-$\#$ symbols is called the *defining length* $\mathcal{L}(H)$ of the schema. Holland obtained a result (the schema theorem) which predicts how the number of strings in a population matching a schema varies from one generation to the next [6]. The theorem[2] is as follows:

$$E[m(H,t+1)] \geq m(H,t) \cdot \underbrace{\frac{f(H,t)}{\bar{f}(t)}}_{Selection} \cdot \underbrace{(1-p_m)^{\mathcal{O}(H)}}_{Mutation} \cdot$$

$$\underbrace{\left[1 - p_c \frac{\mathcal{L}(H)}{N-1} \overbrace{\left(1 - \frac{m(H,t)f(H,t)}{M\,\bar{f}(t)} \right)}^{P_d(H,t)} \right]}_{Crossover} \tag{1}$$

where $m(H,t)$ is the number of strings matching the schema H at generation t, $f(H,t)$ is the mean fitness of the strings in the population matching H, $\bar{f}(t)$ is the mean fitness of the strings in the population, p_m is the probability of mutation per bit, p_c is the probability of one-point crossover per individual, N

---

[1] As pointed out by one of the reviewers, both schemata and strings could be seen as logic formulae, and the problem of matching a string h against a schema H could be formalised using standard substitution techniques used in Logic Programming.

[2] This is a slightly different version of Holland's original theorem (see [3, 21]).

is the number of bits in the strings, M is the number of strings in the population, and $E[m(H, t+1)]$ is the expected value of the number of strings matching the schema H at generation $t+1$. The three horizontal curly brackets beneath the equation indicate which operators are responsible for each term. The bracket above the equation represents the probability of disruption of the schema H at generation t due to crossover, $P_d(H, t)$. Such a probability depends on the frequency of the schema in the mating pool but also on the intrinsic *fragility* of the schema $\frac{\mathcal{L}(H)}{N-1}$.

The representation for schemata based on don't care symbols and binary digits just introduced is not the only one possible. To make it easier for the reader to understand some of the differences between the GP schema definitions introduced in the following sections, in the following we propose (and then modify) a different representation for GA schemata.

A GA schema is fully determined by the defining bits (the 0's and 1's) it contains and by their position. Instead of representing a schema H as a string of characters, one could equivalently represent it with a list of pairs $H = \{(c_1, i_1), ... (c_n, i_n)\}$, where the first element c_j of each pair would represent a group of contiguous defining bits which we call a *component* of the schema, while the second element i_j would represent the position of c_j. For example the schema #10#1 would have two components and could be represented as $\{(10,2),(1,5)\}$.

Although this formal list-of-pairs-based representation is not explicitly used by GA users and researchers, the informal idea of schemata as groups of components is used quite often. For example, when we say that a schema has been disrupted by crossover, we mean that one or more of its components have not been transmitted to the offspring. We do not usually mean that the offspring sample a subspace different from the one represented by the schema (although this is, of course, an entirely equivalent interpretation). Likewise, when we explain the building block hypothesis [3, 6] by saying that GAs work by hierarchically composing relatively fit, short schemata to form complete solutions, we mean that crossover mixes and concatenates the components (defining bits) of low order schemata to form higher order ones. We do not usually mean that crossover is allocating more samples to higher-order schemata representing the intersection of good, lower-order ones.

The schema representation just introduced also seems to be often implicitly assumed when interpreting the GA schema theorem. Obviously, there is no distinction between the number of strings (in the population) sampling the subspace represented by a given schema and the number of times the schema components are jointly present in the strings of the population. Nonetheless, the GA schema theorem is often interpreted as describing the variations of the number of instances of schema components. For example, if $H = $#10#1 the theorem could be interpreted as a lower bound for the expected number of components 10 at position 2 and 1 at position 5 within the population at the next generation.

To reduce the gap between the GA schema theory and some of the GP schema theories introduced in the next section, it is important to understand the effects of altering our component-centred schema representation by omitting the positional information from the pairs. Syntactically this would mean that we represent a schema as a list of groups of bits (components) $H = \{c_1, ..., c_n\}$, like for example {10,1}. Semantically we could interpret a schema as the set containing all the strings which include as substrings the components of the schema (in whatever position). For example, the schema {11,00} would represent all the strings which include both the substring 11 and the substring 00.

An important consequence of removing positional information from the schema definition is that it is possible for a string to include multiple copies of the components of a schema. For example, the components of the schema {10,01} are jointly present four times in the string 10101. This means that with a position-less schema definition there is a distinction between the number of strings sampling the subspace represented by a given schema and the number of times the schema components are jointly present in the strings of the population.

3 GP Schema Theories

While the previous argument might look academic for fixed-length GAs, it is actually very relevant for GP. In some GP schema definitions information about the positions of the schema components is omitted. This has led some researchers to concentrate their analysis on the propagation of such components (often seen as potential building blocks for GP) in the population rather than on the way the number of programs sampling a given schema change over time.

3.1 Theories on Position-less Schema Component Propagation

Given the popularity of Holland's work, Koza [8, pages 116–119] made the first attempt to explain why GP works producing an informal argument showing that Holland's schema theorem would work for GP as well. The argument was based on the idea of defining a schema as the subspace of all trees which contain a pre-defined set of subtrees. According to Koza's definition a schema H is represented as a set of S-expressions. For example the schema $H=\{$(+ 1 x), (* x y)$\}$ represents all programs including at least one occurrence of the expression (+ 1 x) and at least one occurrence of (* x y). This definition of schema was probably suggested by the fact that Koza's GP crossover moves around subtrees. Koza's definition gives only the defining components of a schema not their position, so the same schema can be instantiated (matched) in different ways, and therefore multiple times, in the same program. For example, the schema $H=\{$x$\}$ can be instantiated in two ways in the program (+ x x).

Koza's work on schemata was later formalised and refined by O'Reilly [10, 11] who derived a schema theorem for GP in the presence of fitness-proportionate selection and crossover. The theorem was based on the idea of defining a schema as an unordered collection (a multiset) of subtrees and tree fragments. Tree fragments are trees with at least one leaf that is a "don't care" symbol ('#') which can be matched by any subtree (including subtrees with only one node).

For example the schema $H=\{$(+ # x), (* x y), (* x y)$\}$ represents all the programs including at least one occurrence of the tree fragment (+ # x) and at least *two* occurrences of (* x y).[3] The tree fragment (+ # x) is present in all programs which include a + the second argument of which is x. Like Koza's definition O'Reilly's schema definition gives only the defining components of a schema not their position. So, again the same schema can be instantiated in different ways, and therefore multiple times, in the same program.

O'Reilly's definition of schema allowed her to define the concept of order and defining length for GP schemata. In her definition the order of a schema is the number of non-# nodes in the expressions or fragments contained in the schema. The defining length is the number of links included in the expressions and tree fragments in the schema plus the links which connect them together. Unfortunately, the definition of defining length is complicated by the fact that the components of a schema can be embedded in different ways in different programs. Therefore, the defining length of a schema is not constant but depends on the way a schema is instantiated inside the programs sampling it. As also the total number of links in each tree is variable, this implies that the probability of disruption $P_d(H, h, t)$ of a schema H due to crossover depends on the shape, size and composition of the tree h matching the schema. The schema theorem derived by O'Reilly,

$$E[i(H, t+1)] \geq i(H, t) \cdot \frac{f(H,t)}{\bar{f}(t)} \cdot \left(1 - p_c \cdot \overbrace{\max_{h \in \text{Pop(t)}} P_d(H, h, t)}^{P_d(H,t)} \right), \qquad (2)$$

overcame this problem by considering the maximum of such a probability, $P_d(H, t) = max_{h \in \text{Pop(t)}} P_d(H, h, t)$ which may lead to severely underestimating the number of occurrences of the given schema in the next generation (p_c is the probability of crossover). It is important to note $i(H, t)$ is the number of *instances* of the schema H at generation t and $f(H, t)$ is the mean fitness of the instances of H. This is computed as the weighted sum of the fitnesses of the programs matching H, using as weights the ratios between the number of instances of H each program contains and the total number of instances of H in the population. The theorem describes the way in which the components of the representation of a schema propagate from one generation to the next, rather than the way the number of programs sampling a given schema change during time. O'Reilly discussed the usefulness of her result and argued that the intrinsic variability of $P_d(H, t)$ from generation to generation is one of the major reasons why no hypothesis can be made on the real propagation and use of building blocks (short, low-order relatively fit schemata) in GP. O'Reilly's schema theorem did not include the effects of mutation.

In the framework of his GP system based on context free grammars (CFG-GP) Whigham produced a concept of schema for context-free grammars and the related schema theorem [18, 20, 19]. In CFG-GP programs are the result of applying a set of rewrite rules taken from a pre-defined grammar to a start-

---

[3] We use here the standard notation for multisets, which is slightly different from the one used in O'Reilly's work.

ing symbol S. The process of creation of a program can be represented with a derivation tree whose internal nodes are rewrite rules and whose terminals are the functions and terminals used in the program. In CFG-GP the individuals in the population are derivation trees and the search proceeds using crossover and mutation operators specialised so as to always produce valid derivation trees.

Whigham defines a schema as a partial derivation tree rooted in some non-terminal node, i.e. as a collection of rewrite rules organised into a single derivation tree. Given that the terminals of a schema can be both terminal and non-terminal symbols of a grammar and that the root of a schema can be a symbol different from the starting symbol S, a schema represents all the programs that can be obtained by completing the schema (i.e. by adding other rules to its leaves until only terminal symbols are present) and all the programs represented by schemata which contain it as a component. When the root node of a schema is not S, the schema can occur multiple times in the derivation tree of the same program. This is the result of the absence of positional information in the schema definition. For example, the schema $H = (\text{A} \overset{+}{\Rightarrow} \text{FAA})$ can be instantiated in two ways in the derivation tree, (S (A (F +) (A (F -) (A (T 2)) (A (T x))) (A (T x)))), of the program (+ (- 2 x) x).

Whigham's definition of schema leads to simple equations for the probabilities of disruption of schemata under crossover, $P_{d_c}(H, h, t)$, and mutation, $P_{d_m}(H, h, t)$. Unfortunately as with O'Reilly's, these probabilities vary with the size of the tree h matching the schema. In order to produce a schema theorem for CFG-GP Whigham used the average disruption probabilities of the instances of a schema under crossover and mutation, $\bar{P}_{d_c}(H, t)$ and $\bar{P}_{d_m}(H, t)$, and the average fitness $f(H, t)$ of such instances. The theorem is as follows:

$$E[i(H, t+1)] \geq i(H, t)\frac{f(H, t)}{\bar{f}(t)} \cdot \left\{ [1 - p_m \bar{P}_{d_m}(H, t)] [1 - p_c \bar{P}_{d_c}(H, t)] \right\}, \quad (3)$$

where p_c and p_m are the probabilities of applying crossover and mutation. By changing the grammar used in CFG-GP this theorem can be shown to be applicable both to GAs with fixed length binary strings and to standard GP, of which CFG-GP is a generalisation (see the GP grammar given in [19, page 130]). Like in O'Reilly's case, this theorem describes the way in which the components of the representation of a schema propagate from one generation to the next, rather than the way the number of programs sampling a given schema change over time. The GP schema theorem obtained by Whigham is different from the one obtained by O'Reilly as the concept of schema used by the two authors is different. Whigham's schemata represent derivation-tree fragments which always represent single subexpressions, while O'Reilly's schemata can represent multiple subexpressions.

3.2 Theories on Positioned Schema Propagation

Recently two new schema theories (developed at the same time and independently) have been proposed [17, 14] in which schemata are represented using rooted trees or tree fragments. The rootedness of these schema representations is very important as they reintroduce in the schema definition the positional information lacking in previous definitions of schema for GP. As a consequence a schema can be instantiated at most once within a program and studying the propagation of the components of the schema in the population is equivalent to analysing the way the number of programs sampling the schema change over time.

Rosca [17] has proposed a definition of schema, called *rooted tree-schema*, in which a schema is a rooted contiguous tree fragment. For example, the rooted tree-schema H=(+ # x) represents all the programs whose root node is a + the second argument of which is x. Rosca derived the following schema theorem for GP with standard crossover (when crossover points are selected with a uniform probability):

$$E[m(H,t+1)] \geq m(H,t)\frac{f(H,t)}{\bar{f}(t)} \cdot \tag{4}$$

$$\left[1 - (p_m + p_c) \underbrace{\sum_{h \in H \cap Pop(t)} \overbrace{\frac{\mathcal{O}(H)}{N(h)}}^{P_d(H,h,t)} \frac{f(h)}{\sum_{h \in H \cap Pop(t)} f(h)}}_{P_d(H,t)} \right],$$

where $N(h)$ is the size of a program h matching the schema H, $f(h)$ is its fitness, and the order of a schema $\mathcal{O}(H)$ is the number of defining symbols it contains.[4] Rosca did not give a definition of defining length for a schema. Rosca's schema theorem involves the evaluation of the weighted sum of the fragilities $\frac{\mathcal{O}(H)}{N(h)}$ of the instances of a schema within the population, using as weights the ratios between the fitness of the instances of H and the sum of the fitness of such instances.

In [14] we have proposed a simpler definition of schema for GP which has allowed us to derive a new schema theorem for GP with a new form of crossover, which we call one-point crossover. In the definitions of GP schema mentioned above, in general schemata divide the space of programs into subspaces containing programs of different sizes and shapes. Our definition of schema partitions the program space into subspaces of programs of fixed size and shape.

We define a *schema* as a tree composed of functions from the set $\mathcal{F} \cup \{=\}$ and terminals from the set $\mathcal{T} \cup \{=\}$, where $\mathcal{F}$ and $\mathcal{T}$ are the function set and the terminal set used in a GP run. The symbol $=$ is a "don't care" symbol which stands for a *single* terminal or function. In line with the original definition of schema for GAs, a schema H represents programs having the same shape as H and the same labels for the non-$=$ nodes. For example, if $\mathcal{F}$={+, -} and $\mathcal{T}$={x, y} the schema H=(+ (- = y) =) would represent the four programs (+ (- x y) x), (+ (- x y) y), (+ (- y y) x) and (+ (- y y) y). Our definition of schema is in some sense lower-level than those adopted by others, as a smaller number of trees can be represented by schemata with the same number of "don't care"

---

[4] We have rewritten the theorem in a form which is slightly different from the original one to highlight some of its features.

symbols and it is possible to represent one of their schemata by using a collection of ours.

The number of non-= symbols is called the *order* $\mathcal{O}(H)$ of a schema H, while the total number of nodes in the schema is called the *length* $N(H)$ of the schema. The number of links in the minimum subtree including all the non-= symbols within a schema H is called the *defining length* $\mathcal{L}(H)$ of the schema. For example the schema (+ (- = =) x) has order 3 and defining length 2. These definitions are independent of the shape and size of the programs in the actual population.

In order to derive a GP schema theorem for our schemata we used very simple forms of mutation and crossover, namely point mutation and one-point crossover. *Point mutation* is the substitution of a function in the tree with another function with the same arity or the substitution of a terminal with another terminal. *One-point crossover* works by selecting a common crossover point in the parent programs and then swapping the corresponding subtrees like standard crossover. In order to account for the possible structural diversity of the two parents, one-point crossover starts analysing the two trees from the root nodes and considering for the selection of the crossover point only the parts of the two trees which have the same topology (i.e. the same arity in the nodes encountered traversing the trees from the root node) [14].

The new schema theorem provides the following lower bound for the expected number of individuals sampling a schema H at generation $t+1$ for GP with one-point crossover and point mutation:

$$E[m(H, t+1)] \geq m(H, t) \frac{f(H, t)}{\bar{f}(t)} (1 - p_m)^{\mathcal{O}(H)} \cdot$$

$$\left\{ 1 - p_c \left[p_{\text{diff}}(t) \left(1 - \frac{m(G(H), t) f(G(H), t)}{M \bar{f}(t)} \right) + \right. \right. \tag{5}$$

$$\left. \left. \frac{\mathcal{L}(H)}{(N(H) - 1)} \frac{m(G(H), t) f(G(H), t) - m(H, t) f(H, t)}{M \bar{f}(t)} \right] \right\}$$

where p_m is the mutation probability (per node), $G(H)$ is the zero-th order schema with the same structure of H where all the defining nodes in H have been replaced with "don't care" symbols, M is the number of individuals in the population, $p_{\text{diff}}(t)$ is the conditional probability that H is disrupted by crossover when the second parent has a different shape (i.e. does not sample $G(H)$), and the other symbols have the same meaning as in Equation 2 (see [14] for the proof). The zero-order schemata $G(H)$'s represent different groups of programs all with the same shape and size. For this reason we call them *hyperspaces* of programs. We denote non-zero-order schemata with the term *hyperplanes*.

Equation 5 is more complicated than the corresponding version for GAs [3, 6, 21] because in GP the trees undergoing optimisation have variable size and shape. This is accounted for by the presence of the terms $m(G(H), t)$ and $f(G(H), t)$, which summarise the characteristics of the programs belonging to the same hyperspace in which H is a hyperplane.

In [14] we analysed Equation 5 in detail and discussed the likely interactions between hyperspaces and hyperplanes and the expected variations of the probability $p_{\text{diff}}(t)$ during a run. In particular we conjectured that, given the diversity in the initial population, in general p_{diff} would be quite close to 1 and the probability of schema disruption would be quite high at the beginning of a run. Schema disruption and creation by crossover would then heavily counteract the effects of selection, except for schemata with above average fitness and short defining-length whose shape $G(H)$ is also of above average fitness and is shared by an above average number of programs. We also conjectured that, in the absence of mutation, after a while the population would start converging, like a traditional GA, and the diversity of shapes and sizes would decrease. During this second phase, the competition between schemata belonging to the same hyperspace would become more and more important and the GP schema theorem would asymptotically tend to the standard GA schema theorem. Therefore, during this phase *all* schemata with above average fitness and short defining-length would tend to have a low disruption probability. These conjectures were later corroborated by the experimental study summarised in Sect. 4.

In [11] serious doubts have been cast on the correctness of the building block hypothesis for standard GP. However, our analysis of Equation 5 led us to suggest that it cannot be ruled out in GP with one-point crossover. Nonetheless, as suggested in [5] it is possible that, in order to fully understand what are the building blocks which GP uses to create complete solutions, it will be necessary to study also the propagation of *phenotypical schemata*, i.e. of sets of structurally different but functionally equivalent programs. Unfortunately, it seems very difficult at this stage to imagine how such schemata could be represented and detected.

4 Experimental Studies on GP Schemata

Some researchers have studied the changes over time of quantities which are relevant to schema theorems but which do not require the direct manipulation of all the schemata in a population. For example, Rosca has studied the average of the ratio $f(h)/N(h)$ (an important term in his schema theorem) for all programs in the population and for the best program in the population. However, we will not describe studies of this kind here as they do not explicitly represent and study the schemata in a population but only the programs it contains.

Whatever definition of schema is embraced, it is quite easy to see that the number of different schemata or schema instances contained in a single individual or, worse, in a population of individuals, can be extremely large. For example, using our definition of schema, the number of different schemata contained in a single program h of length $N(h)$ is $2^{N(h)}$, which is large even for short programs. This, together with the complexity of detecting all the instances of a schema with some schema definitions, is probably the reason why only one experimental study on the properties of GP schemata had been done at the time of writing this paper [12]. We describe the main results of this work in the remainder of this section.

In our experiments we have only been able to study (our) schemata in real GP populations by limiting the depth of the programs being evolved to three or four levels and the arity of the functions used by GP to two arguments. Given these restrictions we decided to use in our experiments the XOR problem, which can be solved with these settings. In the experiments we used the function set $\mathcal{F}=\{$AND, OR, NAND, NOR$\}$ and the terminal set $\mathcal{T}=\{$x1, x2$\}$. With these choices and a maximum depth of 2 (the root node is at level 0) we were able to follow the evolution of all the schemata in a population of 50 individuals for 50 generations. (The reader is invited to refer to [12] for a description of other experiments in which we studied subsets of the schemata present in populations of programs with maximum depth 3 and for other details on the experimental setting.)

In a first series of experiments we studied the effects of selection on the propagation of single schemata and on the number of hyperplanes and hyperspace sampled by the programs in the population in ten runs. In each run all schemata in the population were recorded together with: the average fitness of the population in each generation, the average fitness of each schema in each generation, the number of programs sampling the schema in each generation, and the order, length and defining length of the schema. Using these data we were also able to study schema diversity in the population. Diversity was assessed in terms of number of different programs, of schemata and of hyperspaces in each generation.

According to the schema theorem, when selection only is acting, schemata with below average fitness should tend to disappear from the population while above-average schemata should be sampled more frequently. This effect was indeed observed in our runs at the level of both hyperplanes and hyperspaces. However, the results shown in Fig. 1(a) suggest that selection is less effective on hyperspaces than on programs. Indeed, the rate at which hyperspaces disappear is lower than for programs. This can be explained by considering that the average deviation of the fitness of high-order schemata (e.g. programs) from the population fitness is in general bigger than for low-order schemata. Therefore, (positive or negative) selection will be stronger on average for high-order schemata and programs.

Not surprisingly, in none of our runs we observed the exponential schema growth/decay often claimed to happen in GAs [3]. On the contrary, the growth or decay of schemata diverged quite significantly from being monotonic. This was mainly due to the fact that we used small populations of 50 individuals and that genetic drift interfered with the natural growth or decay in the number of schema instances. This effect becomes prominent when the selective pressure decreases, i.e. when nearly all the programs in the population have the same fitness.

In a second set of experiments we considered the effects of normal crossover on the propagation of schemata. As shown in Fig. 1(b), the situation is quite different from the previous case. In an initial phase programs were still subject to a relatively strong selection pressure. As in the selection-only case, hyperspaces were not as strongly affected by it. As soon as the population fitness started saturating the effects of crossover became prevalent. These effects are the constant creation of new individuals and the destruction of old ones. They prevented the population from converging and maintained a high number of different schemata in the population. Towards the end of the runs, there were several times more schemata than in the selection-only case. Standard crossover prevented the competition between hyperspaces from ending as hyperspaces were constantly repopulated.

In a final set of experiments we studied the effects of one-point crossover. One-point crossover behaved quite differently from standard crossover. This behaviour is illustrated by the plots of the population fitness and schema diversity averaged over 10 different runs (Fig. 1(c)). These plots show that the average number of different individuals per hyperspace tended to vary very slowly during the runs. The competition between low-order schemata ended relatively more quickly than the competition between higher order ones. When the competition between hyperspaces was over (not later than generation 19), the one between the programs sampling them was still active. In some cases this took quite a few more generations to finish, but it always did it and, like in the case of selection only, the population always converged. A comparison of the plots in Figures 1(c) and 1(a) reveals that, on average, the rate at which the diversity of high-order schemata changes is reduced by one-point crossover. This is due to the continuous creation of new schemata.

Our experimental study revealed that when the size of the populations used is small genetic drift can be a major component in schema propagation and extinction. This certainly happened in our experiments as soon as the selective pressure decreased, both when selection only was present and when one-point crossover was used. Genetic drift could not become a major driving force in runs with standard crossover, because of the relatively large schema disruption/creation probability associated with it.

Although in the first few generations of runs with standard crossover selection was a primary driving force, the disruption and innovation power of standard crossover remained very strong throughout the entire runs and the population never converged. This contributed to maintaining a certain selective pressure even in late generations, as the average population fitness never reached its maximum.

The relatively large random oscillations of the total number of schemata observed in all runs with standard crossover seem to suggest that, however large, the probability of schema disruption $P_d(H, t)$ in Equation 2 is not constant but may vary significantly from one generation to the next and should be considered a stochastic variable as suggested in [11].

One-point crossover allowed the convergence of the population. So, on average we must assume that its schema disruption probability is smaller than for standard crossover. However, if we compare the total number of schemata in the first few generations of the runs with one-point crossover and the runs with standard crossover, we can see that *initially one-point crossover is as disruptive as standard crossover*. This corroborates our conjecture that schema disruption would be quite high at the beginning of runs with one-point crossover.

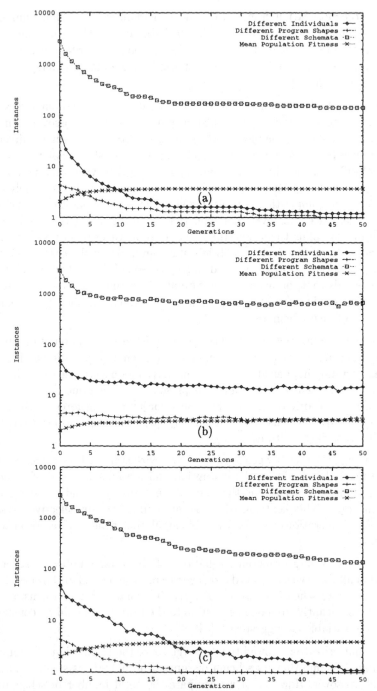

Fig. 1. Average population fitness and schema diversity in 10 GP runs of the XOR problem with: (a) selection only, (b) normal crossover, (c) one-point crossover.

The experiments also corroborate our conjecture that with one-point crossover, after this first highly-disruptive phase in which different hyperspaces compete, the competition between schemata belonging to the same hyperspace becomes the most important effect. In the experiments after generation 19 only one hyperspace was present in all runs. This means that the different programs in such hyperspace had all the same size and shape and that GP with one-point crossover was actually working like a GA with a non-standard crossover. This corroborates our conjecture that the GP schema theorem asymptotically tends to the GA schema theorem.

This behaviour suggests that, while the building block hypothesis seems really in trouble when standard crossover is used, it is as relevant to GP with one-point crossover as it is for traditional GAs.

The convergence properties of GP with one-point crossover suggested that, like in traditional GAs, mutation could be a very important operator to recover lost genetic material. A recent study [13] in which one-point crossover has been compared with standard crossover and an even more constrained form of crossover (called *strict one-point crossover*) has demonstrated that this hypothesis is correct and that point mutation is vital to solve efficiently Boolean classification problems with one-point crossover. (Interestingly, one-point crossover also improves significantly the performance of standard crossover.)

5 Conclusions

In this paper we have reviewed the main results available to date on the theory of schemata for GP, including our own work [14] which is based on a simple definition of the concept of schema for GP which is very close to the original concept of schema in GAs. We have also summarised the first experimental study on GP schemata [12].

At this point the reader might be asking himself or herself: "Is there a *right* schema definition and a *right* schema theory for GP?" In our opinion the answer to this question should be "no", for the following reasons.

We take the viewpoint that a schema is a subspace of the space of possible solutions that schemata are mathematical tools to describe which areas of the search space are sampled by a population. For schemata to be useful in explaining how GP searches, their definition must make the effects of selection, crossover and mutation comprehensible and relatively easy to calculate. A possible problem with some of the earlier definitions of schema for GP is that they make the effects on schemata of the genetic operators used in GP perhaps too difficult to evaluate. However, we believe that any schema definition which, with a particular set of genetic operators, is amenable to mathematical analysis and can lead to some theoretical insight on the inner operation of GP is equally valid and useful.

So, we believe that good different schema theories should not be seen as competitors. When the same operators are used, different schema definitions and schema theorems are simply different views of the same phenomenon. If integrated, they can lead to a deeper understanding. Even more, when it is possible to represent a schema of the particular kind using a set of schemata of another kind it is possible to export some results of the one schema theory to the schemata of another one and *vice versa*. For example, if we consider one of Rosca's schemata, H_R, it is easy to prove that it can be represented using a collection of our schemata, say $H_{PL_1}, \ldots, H_{PL_M}$. Then it is possible to export the results of the Rosca's schema theory to our schemata and our schema theory to Rosca's schemata considering that $E[m(H_R, t)] = \sum_i E[m(H_{PL_i}, t)]$.

More theoretical work will be needed to explore the relations between different theories and the ways they can integrate and cross-fertilise each other. We also believe that more work needs to be done to evaluate the effects of schema creation. This has been one of the main themes of our recent research work, carried out in collaboration with Una-May O'Reilly (MIT AI Lab). This has led to the formulation of new general schema theorems which estimate the mean, the variance and the signal-to-noise ratio of the number of individuals sampling a schema in the presence of schema creation, as well as to the estimation of the short-term probability of extinction of newly created schemata [15].

At the same time, we also think that it will be necessary to produce more empirical studies on the creation and disruption of schemata and on the building block hypothesis. In very recent work we have performed the second study of this kind using our schemata to investigate the possible deceptiveness of the Ant problem [9].

Acknowledgements

The authors wish to thank the members of the EEBIC (Evolutionary and Emergent Behaviour Intelligence and Computation) group for useful discussions and comments, and the reviewers of this paper for their very constructive suggestions. This research is partially supported by a grant under the British Council-MURST/CRUI agreement and by a contract with the Defence Research Agency in Malvern.

References

1. L. Altenberg. The Schema Theorem and Price's Theorem. In L. D. Whitley and M. D. Vose, editors, *Foundations of Genetic Algorithms 3*, pages 23–49, Estes Park, Colorado, USA, 31 July–2 Aug. 1994 1995. Morgan Kaufmann.
2. P. J. Angeline and K. E. Kinnear, Jr., editors. *Advances in Genetic Programming 2*. MIT Press, Cambridge, MA, USA, 1996.
3. D. E. Goldberg. *Genetic Algorithms in Search, Optimization, and Machine Learning*. Addison-Wesley, Reading, Massachusetts, 1989.
4. J. J. Grefenstette. Deception considered harmful. In L. D. Whitley, editor, *Foundations of Genetic Algorithms 2*, San Mateo, CA, 1993. Morgan Kaufman.
5. T. Haynes. Phenotypical building blocks for genetic programming. In E. Goodman, editor, *Genetic Algorithms: Proceedings of the Seventh International Conference*, Michigan State University, East Lansing, MI, USA, 19-23 July 1997. Morgan Kaufmann.

6. J. Holland. *Adaptation in Natural and Artificial Systems*. MIT Press, Cambridge, Massachusetts, second edition, 1992.
7. K. E. Kinnear, Jr., editor. *Advances in Genetic Programming*. MIT Press, Cambridge, MA, 1994.
8. J. R. Koza. *Genetic Programming: On the Programming of Computers by Means of Natural Selection*. MIT Press, 1992.
9. W. B. Langdon and R. Poli. Why ants are hard. Technical Report CSRP-98-4, University of Birmingham, School of Computer Science, January 1998.
10. U.-M. O'Reilly. *An Analysis of Genetic Programming*. PhD thesis, Carleton University, Ottawa-Carleton Institute for Computer Science, Ottawa, Ontario, Canada, 22 Sept. 1995.
11. U.-M. O'Reilly and F. Oppacher. The troubling aspects of a building block hypothesis for genetic programming. In L. D. Whitley and M. D. Vose, editors, *Foundations of Genetic Algorithms 3*, pages 73–88, Estes Park, Colorado, USA, 31 July–2 Aug. 1994 1995. Morgan Kaufmann.
12. R. Poli and W. B. Langdon. An experimental analysis of schema creation, propagation and disruption in genetic programming. In E. Goodman, editor, *Genetic Algorithms: Proceedings of the Seventh International Conference*, Michigan State University, East Lansing, MI, USA, 19-23 July 1997. Morgan Kaufmann.
13. R. Poli and W. B. Langdon. Genetic programming with one-point crossover. In P. K. Chawdhry, R. Roy, and R. K. Pant, editors, *Second On-line World Conference on Soft Computing in Engineering Design and Manufacturing*. Springer-Verlag London, 23-27 June 1997.
14. R. Poli and W. B. Langdon. A new schema theory for genetic programming with one-point crossover and point mutation. In J. R. Koza, K. Deb, M. Dorigo, D. B. Fogel, M. Garzon, H. Iba, and R. L. Riolo, editors, *Genetic Programming 1997: Proceedings of the Second Annual Conference*, pages 278–285, Stanford University, CA, USA, 13-16 July 1997. Morgan Kaufmann.
15. R. Poli, W. B. Langdon, and U.-M. O'Reilly. Short term extinction probability of newly created schemata, and schema variance and signal-to-noise-ratio theorems in the presence of schema creation. Technical Report CSRP-98-6, University of Birmingham, School of Computer Science, January 1998.
16. N. J. Radcliffe. Forma analysis and random respectful recombination. In *Proceedings of the Fourth International Conference on Genetic Algorithms*. Morgan Kaufmann, 1991.
17. J. P. Rosca. Analysis of complexity drift in genetic programming. In J. R. Koza, K. Deb, M. Dorigo, D. B. Fogel, M. Garzon, H. Iba, and R. L. Riolo, editors, *Genetic Programming 1997: Proceedings of the Second Annual Conference*, pages 286–294, Stanford University, CA, USA, 13-16 July 1997. Morgan Kaufmann.
18. P. A. Whigham. A schema theorem for context-free grammars. In *1995 IEEE Conference on Evolutionary Computation*, volume 1, pages 178–181, Perth, Australia, 29 Nov. - 1 Dec. 1995. IEEE Press.
19. P. A. Whigham. *Grammatical Bias for Evolutionary Learning*. PhD thesis, School of Computer Science, University College, University of New South Wales, Australian Defence Force Academy, October 1996.
20. P. A. Whigham. Search bias, language bias, and genetic programming. In J. R. Koza, D. E. Goldberg, D. B. Fogel, and R. L. Riolo, editors, *Genetic Programming 1996: Proceedings of the First Annual Conference*, pages 230–237, Stanford University, CA, USA, 28-31 July 1996. MIT Press.
21. D. Whitley. A genetic algorithm tutorial. Technical Report CS-93-103, Department of Computer Science, Colorado State University, August 1993.

Where Does the Good Stuff Go, and Why?
How Contextual Semantics Influences Program
Structure in Simple Genetic Programming

David E. Goldberg[1] and Una-May O'Reilly[2]

[1] University of Illinois at Urbana-Champaign, IL, USA, 61801
[2] Massachusetts Institute of Technology, MA, USA, 02139

Abstract. Using deliberately designed primitive sets, we investigate the relationship between context-based expression mechanisms and the size, height and density of genetic program trees during the evolutionary process. We show that contextual semantics influence the composition, location and flows of operative code in a program. In detail we analyze these dynamics and discuss the impact of our findings on micro-level descriptions of genetic programming.

1 Introduction

Genetic programmers look at work in genetic algorithms (GAs, [1]) with an odd mixture of scorn and envy. The scorn is born of the knowledge that genetic algorithmists limit themselves—often times needlessly—to somewhat impoverished and sometimes inadequate codings of the problem at hand. The envy is born of the knowledge that genetic programming (GP, [2]) practitioners have less theory than their GA counterparts have to help them size their populations, code and analyze their problems, choose their operator probabilities, and otherwise intelligently make the myriad implementation decisions that every GP practitioner must make.

Our immediate agenda attempts to narrow this gap by developing and testing a theory of spatial-temporal distribution of *expressed* or *operative* genetic material in genetic programming operating on very simple problems. Individually, our theory links spatial structure of program trees to *the expression mechanisms* of a primitive set, and it links temporal structure largely to the *salience* or *fitness structure* of the objective function. The salience-time relationship is studied elsewhere ([3]), and in this contribution we focus solely on the relationship between expression mechanism and spatial structure. We isolate the relationship between what we shall call the *virtual shape* of a program and its expression mechanisms. Virtual shape refers to the structure of a program tree's operative primitives. We borrow our methodology from analogous work in GAs ([4, 5]), starting from the simplest possible problems—problems analogous to the OneMax function of the GA literature ([6]—complicating the problems just enough to enable us to (1) make theoretical predictions of the relationship between expression and spatial structure and (2) empirically test those predictions and be reasonably sure that we are isolating the relevant phenomena.

In the remainder, we start by briefly reviewing related work in the GP literature and discussing the design methodology and decomposition we are borrowing
from GAs. We continue by detailing both the problem we choose to study and
two different expression mechanisms we adopt to idealize real expressive behavior in GP. The connection between those idealized models and real GP is briefly
discussed, and this is followed by both our predictions of outcome and a set of
experiments. The paper concludes with a summary and a list of topics for future
work.

2   Brief Survey of Related GP Literature

The GP community has been interested in program size dynamics for two main
reasons. First, to understand and address bloat ([7, 8, 9, 10, 11]) and second,
to use both bloat and program growth to formulate a micro-level description of
GP search behaviour. Knowledge of program size and growth elucidates the
detailed behaviour of crossover. One can consider where in a program tree
crossover is likely to remove a subtree, and what size, shape and operative nature
the swapped-in or retained subtree is likely to have. These data are correlated
with the quantity of structural change between parent and offspring. Structural
change is in turn crudely correlated with fitness change. This makes it is possible to describe how the algorithm searches towards candidate solutions with
(hopefully) improving fitness.

For example, such a micro-level description can be pieced together from a
series of papers by Rosca and Ballard ([12, 13, 14]). They propose that the
building blocks of GP are the subtrees that extend downwards from the root
of a program tree (i.e. a rooted-tree schema). This rooted subtree provides an
evaluation context for the lower part of the program tree. There are competitions
between schemas of equal order and different shapes or within a schema (same
order and shape). Schemas "grow" downwards extending the ends of a program
tree. In [13], given assumptions of mostly full trees and the often-used 90 : 10
node to leaf crossover point selection bias, with simple calculations it is shown
that the expected height of subtrees swapped by crossover approximately equals
two. In fact, when the population is composed of mostly full trees (which occurs
quickly after maximum height saturation) most crossover branch points are 1/4
as deep as the parent trees and effectively crossover swaps consist of small blocks
of code. Collectively, the search process is a competition among progressively
larger, fitter rooted-tree schemas (growing downward) and exploration of their
combinations with smaller subtrees cut from lower branches.

Soule and Foster in [15] did an interesting experimental study of program size
and depth flows. It was to understand bloat as well as to examine the relative flow
rates of program depth and program size in order to estimate program density.
Their findings have an impact on the usefulness of theories based on assumptions
of full trees such as Rosca's. Based on experimentation using a robot guidance
problem, they state

In addition to the flow towards larger program sizes, there is also motion through the program space towards deeper programs. This flow continues as long as code growth occurs. Overall the programs become less dense during evolution, implying that the flow towards deeper programs occurs more rapidly than the flow towards larger programs. As shown by Equation 3 less dense programs imply that relatively larger sections of code will be involved in the crossover operations, with potentially large repercussions for the evolutionary process. [3] [15]

So, while explanations of behaviour under the assumptions of almost full trees may be correct, these assumptions do not always hold. Currently, there is a gap in current explanations of GP for the cases where programs are not almost full.

We noted that Soule and Foster steered away from relating the semantics of the particular primitive set they were using to the flows they observed. This was quite contrary to our intuition that a theory of program density, size and depth flows should account for the semantics of the primitive set. Primitives in GP programs interact with each other in intricate ways and this results in complicated outcomes in the evolutionary process. Somehow, this factor must relate to spatial structure flow though it may be hidden at the level of operative code flows and only be detectable if they are isolated. In neglecting the role of semantics one assumes, regardless of it, that tree flows are the same for primitive sets with the same syntax. However, semantics influences what is in a program tree and where "it" is. This is important when considering bloat or fitness correlation with crossover subtree swap size. For us the question arose: while size and depth flows are inevitably somewhat very specific to a primitive set and fitness structure of a problem, are there are general properties of a primitive set's expression mechanisms that guide spatial structure?

The central hypothesis of this work is that the context-based expression mechanisms of a primitive set influence virtual shape and this has implications on the evolutionary process. Where the operative, i.e. "good stuff" is in a tree, and how it gets there via crossover and selection influences has implications on the phenomenom of bloat and the fitness optimizing power of GP. It is now timely to review how the GA field pursued an integrated theoretical and empirical methodology. This tack indicates how to answer the GP questions herein.

3 Lessons from GAs: Lessons for GP

Elsewhere ([16, 17]) a design methodology has been articulated for complex conceptual machines such as genetic algorithms and genetic programming. As early as 1990 ([18]), there has been an effective decomposition for the design of

---

[3] One qualification, Soule and Foster's experiments were conducted with very large maximum program depths so they focused on behaviour prior to height (and size) saturation and their experiments selected crossover points uniformly, not with a 90:10 bias.

competent genetic algorithms. Time and time again, experience has shown that it is critical in the design of conceptual machines to walk before you run. Practically speaking, this has meant that it is important to address a linear problem before one addresses a non-linear problem; that it is important to understand a simple expression mechanism before one tackles a more complex one. In this way, lessons are learned about both theory and experimental method in the simpler case, and these lessons can (1) either carry over to the complex case, or (2) not. The latter condition seems like a deal breaker, but understanding that a simple phenomenon does not carry over can be extremely valuable because it tells you what you are not looking at and enables the researcher to search for other answers in a much reduced search space.

In addressing the basic questions of GP performance before the community, researchers have jumped to try to answer the question of "what is a building block?" without understanding the answer to the question "what is an expressed gene?" Of course, in simple GAs, the answer to the latter question is a matter of definition because the presence or absence of a simple allele determines what is expressed. In GP, depending upon the function set, the presence of genetic material may or may not affect the solution—the gene is either *expressed* or not—and if it is expressed, where and with whom it is placed may greatly affect what the gene does—the gene can operate in widely different *contexts*. In the next section, we will try to isolate the notion of expression in an otherwise simple function domain, by constructing two bounding mechanisms of expression. We will argue that these mechanisms are related to those that are often found in real GP in important ways.

4 Two Primitive Sets with Different Expression Mechanisms

To study how different context expression mechanisms relate to spatial structure we will simplify the intricate expression mechanisms of "real" GP problems into a simple mechanism that solely reflects context. Consider two GP primitive sets ORDER and MAJORITY. Let the quantity, arity and names of the primitives in ORDER and MAJORITY be identical. That is, the primitives sets have identical syntax. Each set contains the primitive JOIN of arity two and the complementary primitive pairs $(X_i, \bar{X}_i), i = 0 \ldots n$ which are leafs.

Next, let the semantics of the primitives differ with respect to the particular behaviour two identical program trees exhibit. In the case of MAJORITY, based on the quantity of X_i versus $\bar{X}_i$ in the program tree, only the primitive in greater quantity (majority) is "executed" and output. That is, all instances of the primitive in greater quantity are *operative*. In the case of ORDER, based on the order of X_i versus its complement primitive, only the primitive which appears first when the program elements are inspected inorder is "executed" and output (i.e. operative).

Now, when the same task is required of both primitives sets, for example, output the set of $X_i \, \forall \, i = 0 \ldots n$, the evolved programs that successfully accomplish this task will be different in terms of their program trees but identical in terms of their output.

A well-read GA follower will recognize that the task we cite is the GP version of the standard GA workhorse: the Max-Ones problem (e.g. [19]). We shall use it with a uniform fitness structure. For each different X_i it outputs, a program is awarded a unit. Before proceeding with experiments using ORDER and MAJORITY, we need ask what relation these idealized expression mechanisms have with real GP.

5 The Connection to Real GP

To derive good explanations for GP it is often useful to idealize processes to make them simpler so they may yield to practical theory and experiment, but at the same time it is important to understand the connections between the idealized case and the more complex real world. The GP practitioner, looking at our idealizations might be tempted to criticize them as having little or no relation to the real world, but we believe such a view is incorrect, in the following ways.

In many GP situations, e.g., symbolic regression, parity and multiplexer, practitioners and researchers have noticed that solutions *accumulate* through the addition of more nodes to the tree, thereby allowing a solution to become more and more accurate. The refinement of numerical parameters occurs in this way as does the refinement of various kinds of symbolic models. In a similar way, our MAJORITY mechanism accumulates information about the correct solution, using the aggregate response of large number of variables to give the solution. Thus, we may think of the MAJORITY situation as a "bounding" model of the *accumulative expression* observed in symbolic regression and elsewhere.

At the opposite end of some expression spectrum, the position of a particular function or terminal will (1) prevent the expression of some others, or (2) radically alter how often some others are expressed. Logical functions are examples of the first case. The branch condition of an IF statement directs control over its true and false branches. In the second case, a loop control expression dictates how often the code in the loop body is executed. We will call such expression mechanisms *transformational* because of the radical effect they have on the rest of a program. There is a wide range of transformational expressions. Here we use a simple model of an *eliminative* transformational expression, so called because the presence or absence of a given element in a position relative to some others eliminates these others from consideration. Not all transformational expression mechanisms can be understood through reference to eliminative mechanisms; however, an expression mechanism that completely masks the expresssion of a lot of material seems helpful in understanding the dynamics of bloat and trying to understand where good material may or may not reside in a tres. As such, eliminative expression is a "bounding" model of transformational expression in GP. Furthermore, eliminative expression and accumulative expression are at different ends of the context spectrum.

With these connections to real GP made more explicit, we now turn toward a number of predictions of expected results.

6 Predictions

6.1 The Effect of Selection on Both Expression Mechanisms

Selection is based upon the objective function's assessment of a program's behaviour, rather than its spatial structure. As a control of our study, the same problem using the same objective function is presented to primitive sets which both have sufficient expressive power to formulate a solution. Therefore, we would expect the selection effects to be similar with both expression mechanisms. This prediction can be confirmed by comparing ensemble fitness time series plots for both the best individual (BI) and the population.

For both expression mechanisms, a program is more likely to retain or improve over successive generations if it holds as few complement primitives (i.e. $\bar{X}_i$) as possible. This is because, by simply having a complement primitive, it runs the risk of a crossover operation removing one or more true primitives matching the complementary one (i.e. X_i) and adversely affecting the program's output. Put in another way, selection should eventually "weed" out all the complement primitives in the population (i.e. their quantities should subside to a very low, almost constant level) because they pose a risk in terms of changing some of a program's correct behaviour into incorrect behaviour given an unlucky crossover. We were unable to predict (though we wanted to know) whether the average generation at which the level of complementary primitives in the population drops to a minimal level is the same for both expression mechanisms? And, what is the rate of complementary primitive loss?

In detail, selection pressure should have three successive stages during the evolutionary process. At the beginning, fitter programs are ones that output a greater than average quantity of different X_i's or that express X_i rather than its complement more often than average. After the $\bar{X}_i$ primitives become extinct because they are not fitness positive, fitter programs are ones that output a greater than average quantity of different X_i's. Finally, if the population converges to correct solutions, selection is neutral.

6.2 The MAJORITY Expression Mechanism and Spatial Structure

When considering spatial structure, we have to account for the complicated interaction between selection and crossover. Selection has a first-order fitness effect and "chooses" what programs will be reproduced. Crossover has direct effect on spatial structure because it excises and replaces subtrees. However, it modifies only the trees chosen by selection.

We expect the spatial structure on the initial population of both expression mechanisms to be random within the tolerances of the initial tree height parameter (we use a value of 2). Initially there will be a short transient stage where spatial structure remains random until selection starts to bias population composition.

The only aspect of spatial structure that the MAJORITY expression mechanism directly impacts is program size. After initial population effects subside, during the time interval when there are $\bar{X}_i$'s in present in the population, a program can increase its crossover survival probability by growing bigger in any dimension (becoming deeper or fuller) through increasing its quantities of X_i relative to $\bar{X}_i$. Once the $\bar{X}_i$'s have become extinct, bulking with more X_i's is still a good protective mechanism because it ensures at least a neutral fitness change. Once the population has converged to perfect solutions, bulking with even more X_i's helps to prevent a perfect solution from being destroyed by crossover. Trees will bulk in size anywhere (i.e. grow either downwards or by filling existing levels) given a maximum height which permits downward expansions. After a tree meets the maximum height limit, it will only fill out.

While we can make explicit size predictions, there is no property in the MAJORITY expression mechanism that maps to a specific density flow. This can be explained by the fact that, though the composition of a program depends on the the rate of $\bar{X}_i$ disappearing and the rate of a program bulking up with "extra" X_i's, the *location* of the extra or necessary code is completely left to chance under the expression mechanism. All code of a ORDER program is operative. Correct X_i's can be scattered anywhere in the leafs of a tree. This implies there is no advantageous crossover protection mechanism based on density.

6.3   The ORDER Expression Mechanism and Spatial Structure

With the ORDER expression mechanism, a primitive is operative if it is the first instance of each $(X_i, \bar{X}_i)$ pair in the inorder traversal of its tree. A correct operative primitive is X_i and an incorrect one is $\bar{X}_i$. We shall call the result of traversing an ORDER tree inorder and recording its leaves as encountered, a leaf sequence.[4]

A leaf (sub)sequence has a node cover set. This is the set of nodes which if any member were chosen as a crossover point and the subtree it rooted was excised, the consequence would be that the leaf sequence is disrupted. We call the tree structure linking a leaf sequence and its node cover set, the leaf sequence's crossover fragment. The probability of disruption of a leaf sequence is the size of its crossover fragment divided by the size of the tree.

There are two useful reference points in the leaf sequence an ORDER program tree. See Figure 1 for an illustration of this. The INHIBIT point marks the leftmost point in the leaf sequence after which all primitives to the right are inoperative. To the left of the INHIBIT point there can be incorrect or correct operative primitives and inoperative primitives placed between operative ones. The NONE-WRONG point divides the partition between the start of the leaf sequence and the INHIBIT point. It is the point where to the left lie only correct,

---

[4] We usually think of this sequence as going from left to right. Hence "following" means to the right and "preceding" to the left.

operative primitives or inoperative primitives. Between the NONE-WRONG and INHIBIT points lies correct operative code that is preceded by an incorrect operative primitive, incorrect operative code or inoperative code that comes after an incorrect operative primitive. For example, Figure 1's leaf sequence outputs $X1, X2, \bar{X}3, \bar{X}4, X5$. The NONE-WRONG point is between positions 5 and 6 because $\bar{X}3$ at position 6 is operative but incorrect. To the right of the NONE-WRONG point, $X1$ and $X2$ are both examples of correct operative primitves. All primitives to the right of position 9 are inoperative. Thus, the INHIBIT point is between positions 9 and 10.

$$1 \quad 2 \quad 3 \quad 4 \quad 5 \qquad 6 \quad 7 \quad 8 \quad 9 \qquad 10 \quad 11 \quad 12 \quad 13$$

$$X1, \bar{X}1, \bar{X}1, \bar{X}1, X2, \mid \bar{X}3, \bar{X}4, X3, X5, \parallel X5, X1, X1, \bar{X}5$$

$$\text{NONE-WRONG} \qquad \uparrow \qquad \text{INHIBIT} \uparrow$$

Fig. 1. Reference points in a ORDER expression mechanism leaf sequence. The single vertical bar is the NONE-WRONG point. To its left are correct operative or inoperative primitives. The double vertical bar is the INHIBIT point. Between the two reference points are incorrect operative primitives, correct operative primitives following an operative primitive or inoperative primitives. To the right of the inhibit point, all primitives are non-operative.

There is a key difference between MAJORITY and ORDER in terms of spatial structure dynamics. In contrast to MAJORITY, among a set of ORDER programs with equal fitness, some programs are more likely to survive crossover intact or improved (in terms of fitness) than others. That is, the different sizes of crossover fragments of the un-inhibited leaf sequences in ORDER trees of equal fitness imply that the trees have different crossover survival probabilities. For example, consider program trees A and B of Figure 2 which have 12 element leaf sets and 21 element program trees. Respectively, their leaf sets are:

$$X1, \bar{X}1, \bar{X}3, \mid \bar{X}1, \bar{X}1, \bar{X}1, \bar{X}1, \bar{X}1, X2, X1, \parallel X2, X3$$

and

$$X1, \mid \bar{X}3, X2, \parallel X1, X2, X3, X1, X2, X3, X1, X2, X3$$

The NONE-WRONG point, marked by a single bar, is between position 3 and 4 of A's leaf sequence and between 1 and 2 of B's leaf sequence. Marked by a double bar, the INHIBIT point is between position 9 and 10 of A's leaf sequence and between 2 and 3 of B's leaf sequence. Both programs output $X1, X2, \bar{X}3$ and thus have the same fitness. However, the likelihood that crossover will be guaranteed to non-negatively impact the fitness of each is completely reversed. Crossover will disrupt the crossover fragment of the inoperative primitive sequence (marked by a dotted line) of tree A with probability 5/21 and of tree B with probability

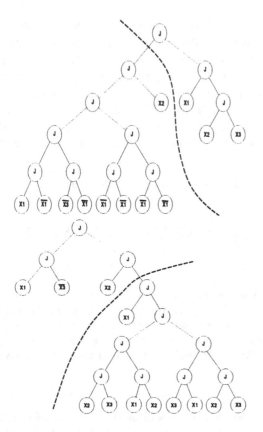

Fig. 2. Both trees output: $X1, X2, \bar{X}3$ but have different likelihood of their inoperative sequences being disrupted.

16/21. Therefore, tree A is much less likely to survive crossover with a neutral or improved fitness than tree B.

The INHIBIT point is helpful in predicting the spatial structure of ORDER trees. It essentially decomposes the leaf sequence into **expressed** (to the left) and **inhibited** (to the right) portions.

Among trees of the same fitness, the trees that have a better chance of surviving crossover intact in terms of their offspring retaining their operative primitives and not being lower in fitness are the ones with relatively large inhibited crossover fragments (i.e. large quantities of (inoperative) material to the right of the INHIBIT point). The more inoperative material a tree has to the right of the INHIBIT point, the more likely it is that some link of the crossover fragment of the inoperative sequence will be chosen as a crossover point. The impact of excising the related subtree will be fitness neutral if the incoming subtree only has primitives that are already expressed in the (original) tree. Or, it will improve fitness if the incoming subtree introduces one or more primitives that become operative in the context of the original tree. Most certainly, change will never decrease fitness. Thus, our expectation is that the inoperative sequence and its crossover fragment will increase in quantity. The shape of the inhibited tree fragment will be completely random as it is composed entirely of inoperative primitives.

In order to improve its crossover survival likelihood, the expressed crossover fragment should be relatively smaller than the inhibited crossover fragment. It should also increase in size but selection dictates that this increase will be from adding inoperative primitives. Excising an inoperative primitive from the operative crossover fragment will not decrease the fitness of the offspring.

To summarize, the entire tree will grow but the inhibited crossover fragment should grow larger faster than the expressed crossover fragment. The entire tree will become full once the tree height limit is reached. The expressed crossover fragment of an almost full tree will be "stringy" reflecting the lower probability of disruption of an expressed crossover fragment.

One would like an understanding of the spatial structure in terms of three partitions (left, between and right of NONE-WRONG and INHIBIT). We found, in fact, this difficult to confidently predict. Instead it will be derived from examining the experiments (see Section 7).

7   Experimental Confirmation

Our runs use a problem size of 16 and the uniform fitness structure of the problem implies that the maximum adjusted fitness is $16 + 1 = 17$. The height of trees in the initial population is 2. Maximum tree height is 15. The population size is 100. Tournament selection is used. Runs proceed for a maximum of 50 generations. Because the distribution of leaf to non-leaf primitives in the primitive set is skewed toward leafs, random tree creation is modified. First a equi-probable random choice is made between using a non-leaf or leaf. Then, if the choice is a leaf, one of the 32 leafs is randomly chosen.

In our simple version of GP, we have GP crossover available with the 90 : 10 biased node to leaf crossover point selection and with non-biased crossover point selection. Using the ORDER expression mechanism we compared the two alternatives with respect to program size, height and density. We measure density in terms of calculating the ratio of (internal) nodes to leafs. A full binary tree will have a node:leaf ratio very close to 1. A fragment could have a node:leaf ratio greater than one because it is not strictly a tree (based on the definition of a node cover set).

Figure 3 has the size, height and node:leaf plots of the crossover comparison. In all three respects there was no significant difference between the crossover point selection alternatives. While we continued to conduct experiments (composed of multiple runs) with both types of crossover, for consistency we present data for runs using the non-biased option. All plotted data is the mean of an ensemble consisting of at least 30 individual GP runs.

7.1   The Effect of Selection

Figure 4 presents the hits times series of the ORDER and MAJORITY expression mechanisms. Note how closely the population hits flows of each resemble each

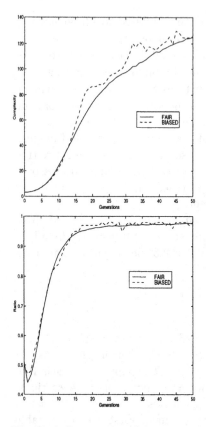

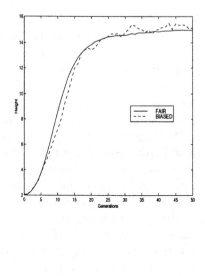

Fig. 3. A Comparison of Crossover Point Selection Options. There is no significant difference in the ensemble mean size, height and node:leaf flows of program trees between selecting a point with the 90:10 bias or fairly. Expression mechanism ORDER was used.

other. While the hits flows of the best individual are slightly less similar, we examined the standard error and found no significant difference between them. This finding confirms the prediction stated in Section 6.1. Because the fitness structure of the problem presented to each expression mechanism is the same, ands the basic semantic power of the the primitive sets is the same, the selection effects should be similar for both expression mechanisms. As a point for future reference, we note that, on average, the best individual attained 15 hits (of a maximum of 16) by generation 15 from a starting quantity of approximately 3.5. The population moves from an initial quantity of 1.5 to slightly less than 15 by generation 30.

Decreasing Levels of Complement Primitives Figure 5 shows that by generation 6, for both expression mechanisms, the fraction of complement primitives in a program, on average, has fallen to 15%. In MAJORITY the fraction continues to drop precipitously so that by generation 15 it is at 2.5% and at generation 25 it

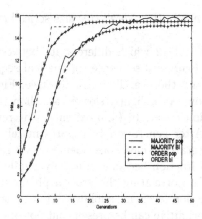

Fig. 4. Hits Time Series.

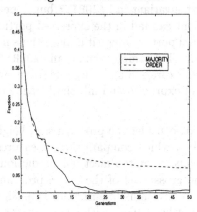

Fig. 5. Fraction of Complement Primitives in a Program (population average).

is below 1%. Subsequently, it remains consistent at 1%. In contrast, ORDER, after generation 6, starts to taper off slowly. It drops an additional 5% (to 10%) after 10 generations and then drops another 5% in the next 35 generations. It settles to a considerably higher consistent level (5% vs 1%) than MAJORITY. Note that this plot data is fractional. The data should be correlated with Figure 6 which shows the program size time series for both expression mechanisms to get an idea of the absolute quantity of complement primitives in a program. In fact, the absolute quantity of complement primitives rises, but the fraction relative to true primitives falls.

The difference of 1% and 5% is worthy of note but we find it somewhat complicated to account for. We can be certain it is not related to selection phenomena because the hits time series for both expression mechanisms are very similar. The same can be stated for a possible relationship to crossover survival mechanisms (i.e. the tendency to bloat) of the entire tree because, per

Figure 6, there is no significant difference in the overall tree size series between both expression mechanisms.

The answer must lie in Figures 9 and 12 which differentiate between the size of the crossover fragment for the expressed sequence (or the leaf sequence preceeding the NONE-WRONG point) and the overall size of the tree. Figure 9 shows that the expressed fragment grows at a slightly slower rate than that of the overall tree but between generation 6 and 16 (approximately) it remains a larger fraction of the overall tree. After generation 16 (approximately), the difference in growth rates subsequently starts to have consequence on the relative size of the expressed fragment and its fraction becomes gradually smaller. The best coherent explanation we can offer correlating the growth phenomena of expressed to total tree size with extinction levels is as follows: First, it is more likely that an individual complement primitive can be present but not operative (i.e. fitness neutral) in ORDER than in MAJORITY in the sense that a complement primitive's expression impact is context invariant in MAJORITY but depends on context in ORDER. In ORDER a complement can fall in the expressed portion and be inoperative or it can be in the inhibited portion where it is also fitness neutral. So overall, ORDER can tolerate a higher level of complements than MAJORITY can. The difference depends on the size of its expressed and inhibited fragments and on the interaction of the growth rates of expressed and inhibited fragments.[5]

ORDER and MAJORITY Compared Next, consider the program size, height and node:leaf ratio plots (Figures 6, 7 and 8) which compare these measures with respect to the two expression mechanisms. They too meet our predictions with respect to the expectation that, at the gross level of the entire program tree, both expression mechanisms would be mostly undistinguishable. The height and size plots shows that size growth does not keep apace with height growth in the first 15 generations. In this interval, the node:leaf ratio increases very steeply after an initial dip. The initial dip is due to the fact that we initialize our population with trees of size 2. It is easier for short binary trees to be full than tall ones because, while the binary arity property implies they must double in size with each height step of 1, in absolute terms, this requires fewer absolute quantities when the height is short. The population has more or less saturated the maximum tree height limit (set to 15) by generation 15. Regardless (of the height limit), size growth continues and, consequently, the node:leaf ratio levels off very close to the absolute possible upper limit of a binary tree. The growth after generation 15 is channelled into fuller trees.

Related to earlier discussion of GP micro-level descriptions, it is worth noting that the major fitness improvements occur before generation 15, i.e. before the tree height limit is reached. In fact, between generations 5 and 10 when the trees fill out in density from 75% to 90%, the population fitness increases by about 30% (5/16). From generations 10 to 15, when the change in density is about

---

[5] Also, it is very compelling to state that complement primitives in ORDER are intron-like. They have non-coding behavior in one context yet coding behaviour in another position on the genotype.

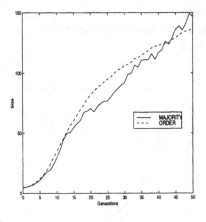

Fig. 6. Size Time Series

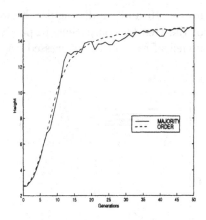

Fig. 7. Height Time Series

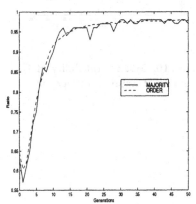

Fig. 8. Node:Leaf Ratio Time Series

5% (from 90% to 95%), the fitness only increases by 20%. As well, many runs find a close to perfect solution by generation 15. Therefore, the activity we are mostly likely to need to understand is in the interval when trees are, intuitively speaking, "stringy" and fitness is (hopefully) improving.

MAJORITY Results Our expectations regarding the spatial structure flows of MAJORITY were coarse: the program trees would get larger but not necessarily in any particular aspect of density. Such expectations hold true. In effect, all leafs of a MAJORITY tree are operative so an examination of Figures 6, 7 and 8 indicating increasing flows imparts the whole story.

ORDER Results Moving on to consider ORDER, the analysis and insights, like our predictions, become more interesting and somewhat more complicated. Figures 9, 10 and 11 distinguish between the expressed and inhibited fragments of the tree by comparing, respectively, the size, probability of disruption and node:leaf ratio of the expressed fragment and entire tree.

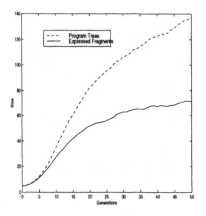

Fig. 9. ORDER: Total Program Size and Expressed Fragment Size.

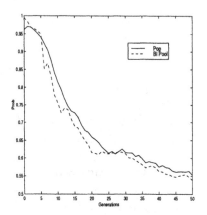

Fig. 10. ORDER: Probability of Disruption of Expressed Fragment.

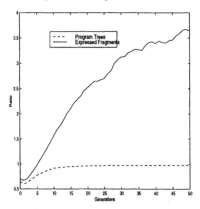

Fig. 11. ORDER: Node:Leaf Ratio Time of Entire Tree and Expressed Fragment.

We discussed the size flow previously in the context of explaining the background level of complement primitives. With respect to the relation between expression mechanism and spatial structure we can note the following: The expressed fragment predominates the program tree in size in early generations (up to generation 6) because growth from a height of 2 to an average height around 7 creates trees which have many more leafs. Selection concurrently favors the trees with more than average *correct* operative primitives. After generation 6 (approximately), both the expressed and inhibited tree fragments grow, however, the overall tree size flow grows more steeply than that of the expressed fragment. The plots of the probability of disruption of the expressed fragment clearly illuminate one of our expectations: that the expressed fragment would appear to protect itself against crossover disruption by exhibiting a decreasing probability of disruption. Figures 9 and 10 indicate this: Despite the expressed code being a very high fraction of the overall program, its probability of disruption decreases. Furthermore, the expressed fragments are "skinny". Their node:leaf ratio exceeds one very quickly and increases with a steep approximately linear slope.

In order to make a finer differentiation, we tracked properties of the crossover fragments associated with the logical partition of the expressed leaf sequence to the left of NONE-WRONG and between NONE-WRONG and INHIBIT. We named the partitions NO-ERRORS and ERRORS-AMONG-CORRECT respectively for obvious reasons. Recall, that we found it difficult to confidently predict these spatial structure flows. Now, with cross-reference to Figures 12 and 13 we can infer three intervals of activity. Prior to generation 6 (when complement primitive levels are rapidly dropping), the relative size of ERRORS-AMONG-CORRECT shrinks because of negative selection pressure. This is based on the fact that ERRORS-AMONG-CORRECT has incorrect operative primitives. The size of NO-ERRORS, which only holds correct or inoperative primitives climbs because of selection pressure which favor programs that express more correct primitives. The relative size of the inhibited fragment (which can be derived from the probability of disruption plot) climbs also, though more slowly than that of NO-ERRORS because its growth is instigated solely as a crossover survival mechanism (this being due to the inhibited fragment's neutral fitness value).

Between generations 6 and 15, the second interval, the complement primitive level is dropping, though more slowly than previously. The relative size of ERRORS-AMONG-CORRECT now decreases more slowly than before because of the consequently weaker negative selection pressure. This is what likely causes the "bulge" in ERRORS-AMONG-CORRECT's size flow during this interval. Both NO-ERRORS and the inhibited fragment increase in relative size because of crossover survival pressure. NO-ERRORS grows more in absolute size because of the fitness pressure which additionally applies to it. However, NO-ERRORS and the inhibited fraction have converse probability of disruption flows. The flow of NO-ERRORS decreases at the approximately inverse steepness of the inhibited flow.

After generation 15, interval 3, a very close to perfect individual has been found. This decreases the selection pressure on NO-ERRORS and its size flow flattens. By generation 30, when most of the population is almost perfectly fit, it flattens even more. The size flow of inhibited, in contrast does not flatten because this fragment is not under the influence of selection pressure due to its fitness neutrality. The probability of disruptions of NO-ERRORS and the inhibited fragments start to converge to 0.5, i.e. equal values.

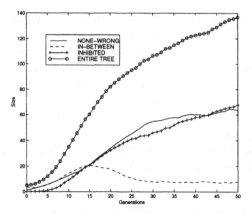

Fig. 12. ORDER: Size Time Series of Three Partitions of a Tree

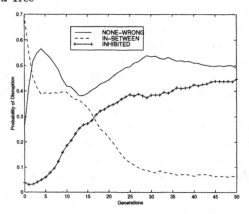

Fig. 13. Probability of Disruption of Three Partitions of a Tree

8 Conclusions

We have asked and started to answer a question important to the analysis and design of genetic programming. Namely, how does the choice of GP primitive set in a particular problem affect the spatial structure of program trees? We believe that answering this question more fully is a key to resolving what appear to be more pressing issues such as understanding and controlling bloat or understanding and promoting the mutation or recombination of effective building blocks. To start answering our more primitive question, we did two important things. First, we adopted a page out of the GA theory handbook and stripped our problem down to the barest of bones, adopting a GP equivalent of a OneMax problem, thereby eliminating problem nonlinearity as a possible concern. Thereafter, we created two idealized expression mechanisms that are in some sense at opposite ends of the contextual spectrum. In the mechanism called ORDER the first oc-

curence of an element turns off all subsequent related elements, whereas in the
MAJORITY mechanism the presence or absence of all related elements may effect
the phenotypic outcome. Although the problem and idealized expression mechanisms are quite simple, we believe that they are good surrogates for expression
mechanisms that are at work in real GP. Specifically, we argued that MAJORITY
represents the kind of accumulative expression that takes place in many practical GP applications such as symbolic regression and the like, and that *order*
represents the eliminative expression that takes place in conditional statements
and other logical program elements, where the presence or absence of an element
eliminates the expression of subsequent related elements.

After creating this idealized framework, we made theoretical predictions of
the expected physical and virtual shape and distribution of the evolved trees.
Here, the word "physical" refers to the presence of genic material whether or not
it is expressed, whereas the term "virtual" concerns itself only with the material that is expressed. Experimental results followed our theoretical predictions
quite well, but the main result to keep in mind is that under accumulative expression, the whole structure was expressed and relevant to the outcome and
under eliminative expression only a long stringy portion of rooted material was
relevant. Understanding these polar extremes of expression in the simple setting
of linear problems should help us understand where the good stuff goes in more
complex idealized problems as well as real GP. To this end, and toward the design of an improved and extended GP that solves large, real world problems, we
recommend a number of investigations in the next section.

9 Future Work

In this paper, we have tried to GP walk before running. In proceeding toward a
GP trot, we recommend the following future work:

- Track the virtual shape flows of simple practical GP problems and observe
  how this correlates with the expression mechanisms in their primitive sets.
- Consider other expression mechanisms or extend the taxonomy of transformational expression mechanisms as defined in this paper.
- Study spatial structure in problems with non-linear fitness structures.
- Investigate the relationship between fitness salience and temporal discovery
  in both independent and nonlinear settings. (For example, see [3]).
- Consider what has been called the decision-making question, particularly
  the relationship between GP population sizing and solution accuracy, in a
  manner akin to GAs ([4, 20]). This will help determine whether GP is just
  a GA with increased variance due to the many more contexts that can exist
  for a code fragment.
- Investigate mixing in easy and hard problems in a manner similar to that of
  GAs ([21, 22, 23]).
- Investigate the design of linkage-promoting recombination operators in a
  manner similar to that of the GA literature ([24, 25, 26]).

In short, what we are proposing here is the development of a facetwise theory for GP. The methodology of using deliberately designed problems, isolating specific properties, and pursuing, in detail, their relationships in simple GP is more than sound; it is the only practical means of systematically extending GP understanding and design. This methodology should lead to GP that works quickly, reliably, and accurately on problems of bounded difficulty. This, in turn, should permit practitioners to apply GP sucessfully day in and day out.

Acknowledgments

U. M. O'Reilly is grateful for funding from the Natural Sciences and Engineering Research Council of Canada and the U.S. Office of Naval Research (contract no. N00014-95-1-1600).

D. E. Goldberg's contribution to this study was sponsored by the Air Force Office of Scientific Research, Air Force Materiel Command, USAF, under grants F49620-94-1-0103, F49620-95-1-0338, and F49620-97-1-0050. The US Government is authorized to reproduce and distribute reprints for Government purposes notwithstanding any copyright notation thereon. The views and conclusions contained herein are those of the authors and should not be interpreted as necessarily representing the official policies or endorsements, either expressed or implied, of the Air Force Office of Scientific Research or the U. S. Government. Professor Goldberg also gratefully acknowledges the use of facilities in the preparation of this paper at the Massacussetts Institute of Technology, the University of Dortmund, and Stanford University during his 1997–98 sabbatical.

References

1. D. E. Goldberg. *Genetic Algorithms in Search, Optimization, and Machine Learning.* Addison-Wesley, Reading, MA, 1989.
2. J. R. Koza. *Genetic Programming: On the Programming of Computers by Means of Natural Selection.* MIT Press, Cambridge, MA, 1992.
3. Una-May O'Reilly and David E. Goldberg. The impact of fitness structure on subsolution acquisition. Technical report, University of Illinois, Urbana-Champaign, Illinois, February 1998. Available via ftp from the Illigal WWW site.
4. D. E. Goldberg, K. Deb, and J. Clark. Genetic algorithms, noise, and the sizing of populations. *Complex Systems,* 6:333–3624, 1992.
5. D. E. Goldberg, B. Korb, and K. Deb. Messy genetic algorithms: Motivation, analysis and first results. *Complex Systems,* 4:415–444, 1989.
6. D. H. Ackley. Stochastic iterated genetic hill-climbing. Technical Report CMU-CS-87-107, Carnegie-Mellon University, Pittsburgh, PA, 1987.
7. Nicholas Freitag McPhee and Justin Darwin Miller. Accurate replication in genetic programming. In L. Eshelman, editor, *Genetic Algorithms: Proceedings of the Sixth International Conference (ICGA95)*, pages 303–309, Pittsburgh, PA, USA, 15-19 July 1995. Morgan Kaufmann.
8. Peter Nordin, Frank Francone, and Wolfgang Banzhaf. Explicitly defined introns and destructive crossover in genetic programming. In Peter J. Angeline and K. E. Kinnear, Jr., editors, *Advances in Genetic Programming 2*, chapter 6, pages 111–134. MIT Press, Cambridge, MA, USA, 1996.

9. W. A. Tackett. *Recombination, Selection, and the Genetic Construction of Computer Programs*. PhD thesis, University of Southern California, Department. of Electrical Engineering Systems, 1994.

10. Justinian P. Rosca and Dana H. Ballard. Complexity drift in evolutionary computation with tree representations. Technical Report NRL5, University of Rochester, Computer Science Department, Rochester, NY, USA, December 1996.

11. David Andre and Astro Teller. A study in program response and the negative effects of introns in genetic programming. In John R. Koza, David E. Goldberg, David B. Fogel, and Rick L. Riolo, editors, *Genetic Programming 1996: Proceedings of the First Annual Conference*, pages 12–20, Stanford University, CA, USA, 28–31 July 1996. MIT Press.

12. Justinian Rosca and Dana H. Ballard. Causality in genetic programming. In L. Eshelman, editor, *Genetic Algorithms: Proceedings of the Sixth International Conference (ICGA95)*, pages 256–263, Pittsburgh, PA, USA, 15-19 July 1995. Morgan Kaufmann.

13. Justinian Rosca. Generality versus size in genetic programming. In John R. Koza, David E. Goldberg, David B. Fogel, and Rick L. Riolo, editors, *Genetic Programming 1996: Proceedings of the First Annual Conference*, pages 381–387, Stanford University, CA, USA, 28–31 July 1996. MIT Press.

14. Justinian Rosca. Analysis of complexity drift in genetic programming. In John R. Koza, Kalyanmoy Deb, Marco Dorigo, David B. Fogel, Max Garzon, Hitoshi Iba, and Rick L. Riolo, editors, *Genetic Programming 1997: Proceedings of the Second Annual Conference*, 28–31 July 1997.

15. Terence Soule and James A. Foster. Code size and depth flows in genetic programming. In John R. Koza, Kalyanmoy Deb, Marco Dorigo, David B. Fogel, Max Garzon, Hitoshi Iba, and Rick L. Riolo, editors, *Genetic Programming 1997: Proceedings of the Second Annual Conference*, pages 313–320, Stanford University, CA, USA, 13-16 July 1997. Morgan Kaufmann.

16. D. E. Goldberg. Making genetic algorithms fly: A lesson from the wright brothers. *Advanced Technology for Developers*, 2:1–8, 1993.

17. D. E. Goldberg. Toward a mechanics of conceptual machines: Developments in theoretical and applied mechanics. In *Proceedings of the Eighteenth Southeastern Conference on Theoretical and Applied Mechanics*, volume 18, 1996.

18. D. E. Goldberg, K. Deb, and B. Korb. Messy genetic algorithms revisited: Studies in mixed size and scale. *Complex Systems*, 4:415–444, 1990.

19. J. Horn and D. E. Goldberg. Genetic algorithm difficulty and the modality of fitness landscapes. In L. D. Whitley and M. D. Vose, editors, *Foundations of Genetic Algorithms*, volume 3, San Mateo, CA, 1995. Morgan Kaufmann. (To appear).

20. G. Harik, E. Cantuú-Paz, D.E. Goldberg, and B.L. Miller. The gambler's ruin problem, genetic algorithms, and the sizing of populations.

21. D. E. Goldberg, K. Deb, and D. Thierens. Toward a better understanding of mixing in genetic algorithms. *Journal of the Society of Instrument and Control Engineers*, 32(1):10–16, 1993.

22. D. Thierens and D. E. Goldberg. Mixing in genetic algorithms. In S. Forrest, editor, *Proceedings of the Fifth International Conference on Genetic Algorithms*, San Mateo, CA, 1993. Morgan Kaufmann.

23. D. Thierens. *Analysis and design of genetic algorithms*. PhD thesis, Katholieke Universiteit Leuven, 1995.

24. D. E. Goldberg, K. Deb, H. Kargupta, and G. Harik. Rapid, accurate optimization of difficult problems using fast messy genetic algorithms. In S. Forrest, editor,

Proceedings of the Fifth International Conference on Genetic Algorithms, pages 56–64, San Mateo, CA, 1993. Morgan Kaufmann.

25. G. Harik. *Learning gene linkage to efficiently solve problems of bounded difficulty using genetic algorithms*. PhD thesis, University of Michigan, Ann Arbor, 1997. Unpublished doctoral dissertation (also IlliGAL 97005).

26. H. Kargupta. Search, evolution, and the gene expression messy genetic algorithm. Technical Report Unclassified Report LA-UR 96-60, Los Alamos National Laboratory, Los Alamos, NM, 1996.

Fitness Causes Bloat: Mutation

W. B. Langdon and R. Poli

School of Computer Science, University of Birmingham, Birmingham B15 2TT, UK
{W.B.Langdon,R.Poli}@cs.bham.ac.uk http://www.cs.bham.ac.uk/~wbl, ~rmp
Tel: +44 (0) 121 414 4791, Fax: +44 (0) 121 414 4281

Abstract. The problem of evolving, using mutation, an artificial ant to follow the Santa Fe trail is used to study the well known genetic programming feature of growth in solution length. Known variously as "bloat", "fluff" and increasing "structural complexity", this is often described in terms of increasing "redundancy" in the code caused by "introns".

Comparison between runs with and without fitness selection pressure, backed by Price's Theorem, shows the tendency for solutions to grow in size is caused by fitness based selection. We argue that such growth is inherent in using a fixed evaluation function with a discrete but variable length representation. With simple static evaluation search converges to mainly finding trial solutions with the same fitness as existing trial solutions. In general variable length allows many more long representations of a given solution than short ones. Thus in search (without a length bias) we expect longer representations to occur more often and so representation length to tend to increase. I.e. fitness based selection leads to bloat.

1 Introduction

The tendency for programs in genetic programming (GP) populations to grow in length has been widely reported [Tac93; Tac94; Ang94; Tac95; Lan95; NB95; SFD96]. This tendency has gone under various names such as "bloat", "fluff" and increasing "structural complexity". The principal explanation advanced for bloat has been the growth of "introns" or "redundancy", i.e. code which has no effect on the operation of the program which contains it. ([WL96] contains a survey of recent research in biology on "introns"). Such introns are said to protect the program containing them from crossover [BT94; Bli96; NFB95; NFB96]. [MM95] presents an analysis of some simple GP problems designed to investigate bloat. This shows that, with some function sets, longer programs can "replicate" more "accurately" when using crossover. I.e. offspring produced by crossover between longer programs are more likely to behave as their parents than children of shorter programs. [PL98] argues the fraction of genetic material changed by crossover is smaller in longer programs. Such local changes may lead to GP populations becoming trapped at local peaks in the fitness landscapes. [RB96a] provides an analysis of bloat using tree schemata specifically for GP.

We advance a more general explanation which should apply generally to any discrete variable length representation and generally to any progressive search technique. That is bloat is not specific to genetic programming applied to trees using tree based crossover but should also be found with other genetic operators and non-population based stochastic search techniques such as simulated annealing and stochastic iterated hill climbing ([Lan98b] contains examples where such bloat does indeed occur and [LP98a] investigates bloat in GP with dynamic fitness functions).

The next section summarises our argument that bloat is inherent in variable length representations such as GP [LP97b]. In Sects. 3 and 4 we expand our previous analysis of a typical GP demonstration problem to include solution by mutation in place of crossover, again showing that it suffers from bloat and also showing that bloat is not present in the absence of fitness based selection. (This improves earlier experiments [LP97c] by removing the arbitrary limit on program size). Section 5 describes the results we have achieved and this is followed in Sect. 6 by a discussion of the potential advantages and disadvantages of bloat and possible responses to it. Finally Sect. 7 summarises our conclusions.

2   Bloat in Variable Length Representations

In general with variable length discrete representations there are multiple ways of representing a given behaviour. If the evaluation function is static and concerned only with the quality of each trial solution and not with its representation then all these representations have equal worth. If the search strategy were unbiased, each of these would be equally likely to be found. In general there are many more long ways to represent a specific behaviour than short representations of the same behaviour. Thus we would expect a predominance of long representations.

Practical search techniques are biased. There are two common forms of bias when using variable length representations. Firstly search techniques often commence with simple (i.e. short) representations, i.e. they have an in built bias in favour of short representations. Secondly they have a bias in favour of continuing the search from previously discovered high fitness representations and retaining them as points for future search. I.e. there is a bias in favour of representations that do at least as well as their initiating point(s).

On problems of interest, finding improved solutions is relatively easy initially but becomes increasingly more difficult. In these circumstances, especially with a discrete fitness function, there is little chance of finding a representation that does better than the representation(s) from which it was created. (Cf. "death of crossover" [Lan98a, page 206]. Section 5.6 shows this holds in this example). So the selection bias favours representations which have the same fitness as those from which they were created.

In general the easiest way to create one representation from another and retain the same fitness is for the new representation to represent identical behaviour. Thus, in the absence of improved solutions, the search may become a random search for new representations of the best solution found so far. As we said above, there are many more long representations than short ones for the same solution, so such a random search (other things being equal) will find more long representations than short ones. In GP this has become known as bloat.

3 The Artificial Ant Problem

The artificial ant problem is described in [Koz92, pages 147–155]. It is a well studied problem and was chosen as it has a simple fitness function. [LP98b] shows its simpler solutions have characteristics often associated with real world programs but that GP and other search techniques find it difficult (possibly due to bloat). Briefly the problem is to devise a program which can successfully navigate an artificial ant along a twisting trail on a square 32 × 32 toroidal grid. The program can use three operations, Move, Right and Left, to move the ant forward one square, turn to the right or turn to the left. Each of these operations takes one time unit. The sensing function IfFoodAhead looks into the square the ant is currently facing and then executes one of its two arguments depending upon whether that square contains food or is empty. Two other functions, Prog2 and Prog3, are provided. These take two and three arguments respectively which are executed in sequence.

The artificial ant must follow the "Santa Fe trail", which consists of 144 squares with 21 turns. There are 89 food units distributed non-uniformly along it. Each time the ant enters a square containing food the ant eats it. The amount of food eaten is used as the fitness measure of the control program.

The evolutionary system we use is identical to [LP97b] except the crossover operator is replaced by mutation. The details are given in Table 1, parameters not shown are as [Koz94, page 655]. On each version of the problem 50 independent runs were conducted. Note in these experiments we allow the evolved programs to be far bigger than required to solve the problem. (The smallest solutions comprise only 11 node [LP98b]).

Table 1. Ant Problem

Objective:	Find an ant that follows the "Santa Fe trail"
Terminal set:	Left, Right, Move
Functions set:	IfFoodAhead, Prog2, Prog3
Fitness cases:	The Santa Fe trail
Fitness:	Food eaten
Selection:	Tournament group size of 7, non-elitist, generational
Wrapper:	Program repeatedly executed for 600 time steps.
Population Size:	500
Max program size:	32,767
Initial population:	Created using "ramped half-and-half" with a max depth of 6
Parameters:	90% mutation, 10% reproduction
Termination:	Maximum number of generations G = 50

4 Tree Mutation

For our purposes it is necessary that the mutation operator be able to change the size of the chromosomes it operates. Ideally this should be unbiased in the sense of producing, of itself, no net change in size. The following operator, in all cases examined, produces offspring which are on average almost the same size as their parent (cf. Sect. 5.7). (Note this may not be the case with different ratios of node branching factors in either the terminal/function set or in the evolving populations). An alternative tree mutation operator which also tries to avoid size bias by creating random trees of a randomly chosen size is proposed in [Lan98b].

The mutation operator selects uniformly at random a node (which may be a function or terminal) and replaces it and the subtree descending from it with a randomly created tree. The new tree is created using the same "ramped half-and-half" method used to create the initial population [Koz92, Page 92–93] however its maximum height is chosen at random from one to the height of the tree it is to replace. Note a terminal will always be replaced by a randomly selected terminal (i.e. there is no change in size) but a function can be replaced by a larger (but not deeper) or smaller tree, so that the program's size may change.

5 Results

5.1 Standard Runs

In 50 independent runs 9 found "ants" that could eat all the food on the Santa Fe trail within 600 time steps. The evolution of maximum and mean fitness averaged across all 50 runs is given by the upper curves in Fig. 1. (In all cases the average minimum fitness is near zero). These curves show the fitness behaving as with crossover with both the maximum and average fitness rising rapidly initially but then rising more slowly later in the runs. The population converges in the sense that the average fitness approaches the maximum fitness. However the spread of fitness values of the children produced in each generation remains large and children which eat either no food or only one food unit are still produced even in the last generation.

Figure 2 shows the evolution of maximum and mean program size averaged across all 50 runs. Although on average mutation runs show somewhat different behaviour of program length compared to crossover they are similar in that after an initial period bloat starts and program lengths grow indefinitely.

5.2 No Selection

A further 50 runs were conducted using the same initial populations and no fitness selection. As with crossover no run found a solution and the maximum, mean and other fitness statistics fluctuate a little but are essentially unchanged In contrast to the case with crossover only, our mutation operator has a slight bias. In the first 50 generation this causes the population to gradually increase in size at the rate of $\frac{1}{3}$ of a node per generation

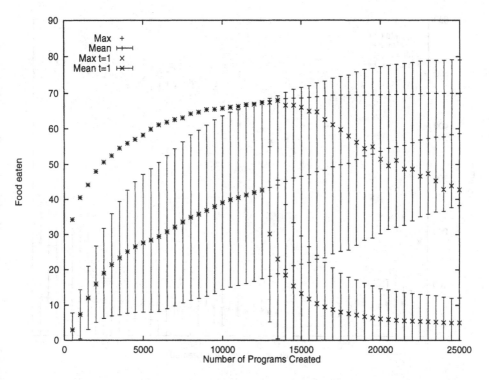

Fig. 1. Evolution of maximum and population mean of food eaten. t=1 curves show effect of removing fitness selection after generation 25. Error bars indicate one standard deviation. Means of 50 runs.

5.3   Removing Selection

A final 50 runs were conducted in which fitness selection was removed after generation 25 (i.e. these runs are identical to those in Sect. 5.1 up to generation 25, 12,500 programs created). The evolution of maximum and mean fitness averaged across all 50 runs is given in Fig. 1. As expected Fig. 1 shows in the absence of fitness selection the fitness of the population quickly falls.

After fitness selection is removed, the length of programs behaves much as when there is no selection from the start of the run. I.e. bloat is replaced by the slow growth associated with the mutation operator and the spread of program lengths gradually increases (cf. Fig. 2)

5.4   Fitness is Necessary for Bloat – Price's Theorem Applied to Representation Size

Price's Covariance and Selection Theorem [Pri70] from population genetics relates the expected change in frequency of a gene Δq in a population from one generation to the next, to the covariance of the gene's frequency in the original population with the number of offspring z produced by individuals in that population:

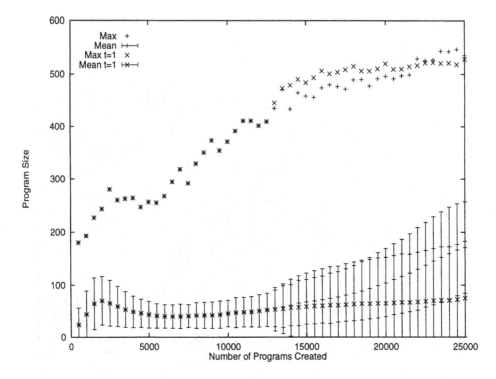

Fig. 2. Evolution of maximum and population mean program length. t=1 curves show effect of removing fitness selection after generation 25. Error bars indicate one standard deviation. Solid line is the length of the "best" program in the population. Means of 50 runs.

$$\Delta q = \frac{\mathrm{Cov}(z, q)}{\overline{z}} \qquad (1)$$

We have used it to help explain the evolution of the number of copies of functions and terminals in GP populations [LP97a; Lan98a]. In our experiments the size of the population does not change so $\overline{z} = 1$ and the expected number of children is given by the parent's rank so in large populations the expected change is approximately $\mathrm{Cov}(t(r/p)^{t-1}, q)$ as long as crossover is random. (t is the tournament size and r is each program's rank within the population of size p).

Where representation length is inherited, such as in GP and other search techniques, (1) should hold for representation length. More formally Price's theorem applies (provided length and genetic operators are uncorrelated) since representation length is a *measurement function* of the genotype [Alt95, page 28].

Where fitness selection is not used (as in the previous sections), each individual in the population has an equal chance of producing children and so the covariance is always zero. Therefore Price's Theorem predicts on average there will be no change in length.

5.5 Correlation of Fitness and Program Size

In all cases there is a positive correlation between program score and length of programs. In the initial population the correlation is .37 on average, indicating that long random programs do better than short ones. This may be because they are more likely to contain useful primitives (such as Move) than short programs. Also short ones make fewer moves before they are reinitialised and re-executed. This may increase the chance of them falling into unproductive short cyclic behaviour that longer, more random, programs can avoid. Selection quickly drives the population towards these better individuals, so reducing the correlation to .13 In the absence of selection (or after selection has been removed halfway through a run) mutation tends to randomise the population so that the correlation remains (or tends towards) that in the initial population.

5.6 Effect of Mutation on Fitness

In the initial generations mutation is disruptive with 64.6% producing a child with a different score from its parent. However mutation evolves, like crossover, to become less disruptive. By the end of the run only 25.4% of mutants have a different score from their parent.

The range of change of fitness is highly asymmetric; many more children are produced which are worse than their parent than those that are better. By the end of the run, only 0.02% of the population are fitter than their parent. Similar behaviour has been reported using crossover on other problems [NFB96] [RB96b, page 183] [Lan98a, Chapter 8].

5.7 Mutation and Program Length

We see from Fig. 3 initially on average mutation makes little change to the length of programs, however, after bloat becomes established, our mutation operator produces children which are slightly shorter than their parents on average. I.e. once the change in the fitness of the population slows, program size bloats despite length changes introduced by mutation. The positive covariance of fitness and length shows this bloat is driven by fitness selection.

6 Discussion

6.1 Do we Want to Prevent Bloat?

From a practical point of view the machine resources consumed by any system which suffers from bloat will prevent extended operation of that system. However in practice we may not wish to operate the system continually. For example it may quickly find a satisfactory solution or better performance may be achieved by cutting short its operation and running it repeatedly with different starting configurations [Koz92, page 758].

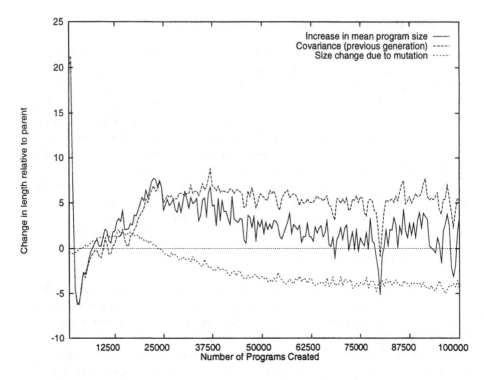

Fig. 3. Mean change in length of offspring relative to parent: normal runs. Means of 50 runs.

In some data fitting problems growth in solution size may be indicative of "over fitting", i.e. better matching on the test data but at the expense of general performance. For example [Tac93, page 309] suggests "parsimony may be an important factor not for 'aesthetic' reasons or ease of analysis, but because of a more direct relationship to fitness: there is a bound on the 'appropriate size' of solution tree for a given problem".

By providing a "defence against crossover" [NFB96, page 118] bloat causes the production of many programs of identical performance. These can consume the bulk of the available machine resources and by "clogging up" the population may prevent GP from effectively searching for better programs.

On the other hand [Ang94, page 84] quotes results from fixed length GAs in favour of representations which include introns, to argue we should "not ... impede this emergent property [i.e. introns] as it may be crucial to the successful development of genetic programs". Introns may be important as "hiding places" where genetic material can be protected from the current effects of selection and so retained in the population. This may be especially important where fitness criteria are dynamic. A change in circumstance may make it advantageous to execute genetic material which had previously been hidden in an intron. [Hay96] shows an example where a difficult GP representation is improved by deliberately inserting duplicates of evolved code.

In complex problems it may not be possible to test every solution on every aspect of the problem and some form of dynamic selection of test cases may be required [GR94]. For example in some cases co-evolution has been claimed to be beneficial to GP. If the fitness function is sufficiently dynamic, will there still be an advantage for a child in performing identically to its parents? If not, will we still see such explosive bloat?

6.2 Three Ways to Control Bloat

Three methods of controlling bloat have been suggested. Firstly, and most widely used is to place a universal upper bound either on tree depth [Koz92] or program length. ([GR96; LP97a] discuss unexpected problems with this approach).

The second (also commonly used) is to incorporate program size directly into the fitness measure (often called parsimony pressure) [Koz92; ZM93; IdS94]. [RB96a] gives an analysis of the effect of parsimony pressure which varies linearly with program length. Multi-objective fitness measures where one objective is compact or fast programs have also been used [Lan96].

The third method is to tailor the genetic operations. [Sim93, page 469] uses several mutation operators but adjusts their frequencies so a "decrease in complexity is slightly more probable than an increase". [Bli96] suggests targeting genetic operations at redundant code. This is seldom used, perhaps due to the complexity of identifying redundant code. [SFD96] showed bloat continuing despite their targeted genetic operations. Possibly this was because of the difficulty of reliably detecting introns. I.e. there was a route whereby the GP could evolve junk code which masqueraded as being useful and thereby protected itself from removal. While [RB96a] propose a method where the likelihood of potentially disruptive genetic operations increases with parent size.

7 Conclusions

We have generalised existing explanations for the widely observed growth in GP program size with successive generations (*bloat*) to give a simple statistical argument which should be generally applicable both to GP and other systems using discrete variable length representations and static evaluation functions. Briefly, in general simple static evaluation functions quickly drive search to converge, in the sense of concentrating the search on trial solutions with the same fitness as previously found trial solutions. In general variable length allows many more long representations of a given solution than short ones of the same solution. Thus (in the absence of a parsimony bias) we expect longer representations to occur more often and so representation length to tend to increase. I.e. current simple fitness based selection techniques lead to bloat.

In earlier work [LP97b] we took a typical GP problem and demonstrated with fitness selection it suffers from bloat whereas without selection it does not. In Sects. 3, 4 and 5 we repeated these experiments replacing crossover with mutation and showed fitness selection can still cause bloat. ([Lan98b] shows bloat can also occur with simulated annealing and hill climbing). We have demonstrated that if fitness selection is removed, bloat is stopped and program size changes little on average. As expected in the absence of selection, mutation is free to change program size at random and the range of sizes increases, as does the mean size (albeit very slowly). Detailed measurement of mutation confirms after an extended period of evolution, most are not disruptive (i.e. most children have the same fitness as their parents).

In Sect. 5.4 we applied Price's Theorem to program lengths within evolving populations. We confirmed experimentally that it fits unless bias in the genetic operators have significant impact. We used Price's Theorem to argue fitness selection is required for a change in average representation length. In Sect. 6 we discussed the circumstances in which we need to control bloat and current mechanisms which do control it but suggest a way forward may be to consider more complex dynamic fitness functions. Preliminary investigations in this direction are reported in [LP98a].

Acknowledgements

This research was funded by the Defence Research Agency in Malvern.

References

[Alt95]   Lee Altenberg. The Schema Theorem and Price's Theorem. In L. Darrell Whitley and Michael D. Vose, editors, *Foundations of Genetic Algorithms 3*, pages 23–49, Estes Park, Colorado, USA, 31 July–2 August 1994 1995. Morgan Kaufmann.

[Ang94]   Peter John Angeline. Genetic programming and emergent intelligence. In Kenneth E. Kinnear, Jr., editor, *Advances in Genetic Programming*, chapter 4, pages 75–98. MIT Press, 1994.

[Bli96]   Tobias Blickle. *Theory of Evolutionary Algorithms and Application to System Synthesis*. PhD thesis, Swiss Federal Institute of Technology, Zurich, November 1996.

[BT94]   Tobias Blickle and Lothar Thiele. Genetic programming and redundancy. In J. Hopf, editor, *Genetic Algorithms within the Framework of Evolutionary Computation (Workshop at KI-94, Saarbrücken)*, pages 33–38, Im Stadtwald, Building 44, D-66123 Saarbrücken, Germany, 1994. Max-Planck-Institut für Informatik (MPI-I-94-241).

[GR94]   Chris Gathercole and Peter Ross. Dynamic training subset selection for supervised learning in genetic programming. In Yuval Davidor, Hans-Paul Schwefel, and Reinhard Männer, editors, *Parallel Problem Solving from Nature III*, pages 312–321, Jerusalem, 9-14 October 1994. Springer-Verlag.

[GR96]   Chris Gathercole and Peter Ross. An adverse interaction between crossover and restricted tree depth in genetic programming. In John R. Koza, David E. Goldberg, David B. Fogel, and Rick L. Riolo, editors, *Genetic Programming 1996: Proceedings of the First Annual Conference*, pages 291–296, Stanford University, CA, USA, 28–31 July 1996. MIT Press.

[Hay96] Thomas Haynes. Duplication of coding segments in genetic programming. In *Proceedings of the Thirteenth National Conference on Artificial Intelligence*, pages 344–349, Portland, OR, August 1996.

[IdS94] Hitoshi Iba, Hugo de Garis, and Taisuke Sato. Genetic programming using a minimum description length principle. In Kenneth E. Kinnear, Jr., editor, *Advances in Genetic Programming*, chapter 12, pages 265–284. MIT Press, 1994.

[Koz92] John R. Koza. *Genetic Programming: On the Programming of Computers by Natural Selection*. MIT Press, Cambridge, MA, USA, 1992.

[Koz94] John R. Koza. *Genetic Programming II: Automatic Discovery of Reusable Programs*. MIT Press, Cambridge Massachusetts, May 1994.

[Lan95] W. B. Langdon. Evolving data structures using genetic programming. In L. Eshelman, editor, *Genetic Algorithms: Proceedings of the Sixth International Conference (ICGA95)*, pages 295–302, Pittsburgh, PA, USA, 15-19 July 1995. Morgan Kaufmann.

[Lan96] William B. Langdon. Data structures and genetic programming. In Peter J. Angeline and K. E. Kinnear, Jr., editors, *Advances in Genetic Programming 2*, chapter 20, pages 395–414. MIT Press, Cambridge, MA, USA, 1996.

[Lan98a] W. B. Langdon. *Data Structures and Genetic Programming*. Kulwer, 1998. Forthcoming.

[Lan98b] W. B. Langdon. The evolution of size in variable length representations. In *1998 IEEE International Conference on Evolutionary Computation*, Anchorage, Alaska, USA, 5-9 May 1998. Forthcoming.

[LP97a] W. B. Langdon and R. Poli. An analysis of the MAX problem in genetic programming. In John R. Koza, Kalyanmoy Deb, Marco Dorigo, David B. Fogel, Max Garzon, Hitoshi Iba, and Rick L. Riolo, editors, *Genetic Programming 1997: Proceedings of the Second Annual Conference*, pages 222–230, Stanford University, CA, USA, 13-16 July 1997. Morgan Kaufmann.

[LP97b] W. B. Langdon and R. Poli. Fitness causes bloat. In P. K. Chawdhry, R. Roy, and R. K. Pan, editors, *Second On-line World Conference on Soft Computing in Engineering Design and Manufacturing*. Springer-Verlag London, 23-27 June 1997.

[LP97c] W. B. Langdon and R. Poli. Fitness causes bloat: Mutation. In John Koza, editor, *Late Breaking Papers at the GP-97 Conference*, pages 132–140, Stanford, CA, USA, 13-16 July 1997. Stanford Bookstore.

[LP98a] W. B. Langdon and R. Poli. Genetic programming bloat with dynamic fitness. In W. Banzhaf, R. Poli, M. Schoenauer, and T. C. Fogarty, editors, *Proceedings of the First European Workshop on Genetic Programming*, LNCS, Paris, 14-15 April 1998. Springer-Verlag. Forthcoming.

[LP98b] W. B. Langdon and R. Poli. Why ants are hard. Technical Report CSRP-98-4, University of Birmingham, School of Computer Science, January 1998. submitted to GP-98.

[MM95] Nicholas Freitag McPhee and Justin Darwin Miller. Accurate replication in genetic programming. In L. Eshelman, editor, *Genetic Algorithms: Proceedings of the Sixth International Conference (ICGA95)*, pages 303–309, Pittsburgh, PA, USA, 15-19 July 1995. Morgan Kaufmann.

[NB95] Peter Nordin and Wolfgang Banzhaf. Complexity compression and evolution. In L. Eshelman, editor, *Genetic Algorithms: Proceedings of the Sixth International Conference (ICGA95)*, pages 310–317, Pittsburgh, PA, USA, 15-19 July 1995. Morgan Kaufmann.

[NFB95] Peter Nordin, Frank Francone, and Wolfgang Banzhaf. Explicitly defined introns and destructive crossover in genetic programming. In Justinian P. Rosca, editor, *Proceedings of the Workshop on Genetic Programming: From Theory to Real-World Applications*, pages 6–22, Tahoe City, California, USA, 9 July 1995.

[NFB96] Peter Nordin, Frank Francone, and Wolfgang Banzhaf. Explicitly defined introns and destructive crossover in genetic programming. In Peter J. Angeline and K. E. Kinnear, Jr., editors, *Advances in Genetic Programming 2*, chapter 6, pages 111–134. MIT Press, Cambridge, MA, USA, 1996.

[PL98] Riccardo Poli and William B Langdon. On the ability to search the space of programs of standard, one-point and uniform crossover in genetic programming. Technical Report CSRP-98-7, University of Birmingham, School of Computer Science, January 1998. submitted to GP-98.

[Pri70] George R. Price. Selection and covariance. *Nature*, 227, August 1:520–521, 1970.

[RB96a] Justinian P. Rosca and Dana H. Ballard. Complexity drift in evolutionary computation with tree representations. Technical Report NRL5, University of Rochester, Computer Science Department, Rochester, NY, USA, December 1996.

[RB96b] Justinian P. Rosca and Dana H. Ballard. Discovery of subroutines in genetic programming. In Peter J. Angeline and K. E. Kinnear, Jr., editors, *Advances in Genetic Programming 2*, chapter 9, pages 177–202. MIT Press, Cambridge, MA, USA, 1996.

[SFD96] Terence Soule, James A. Foster, and John Dickinson. Code growth in genetic programming. In John R. Koza, David E. Goldberg, David B. Fogel, and Rick L. Riolo, editors, *Genetic Programming 1996: Proceedings of the First Annual Conference*, pages 215–223, Stanford University, CA, USA, 28–31 July 1996. MIT Press.

[Sim93] K. Sims. Interactive evolution of equations for procedural models. *The Visual Computer*, 9:466–476, 1993.

[Tac93] Walter Alden Tackett. Genetic programming for feature discovery and image discrimination. In Stephanie Forrest, editor, *Proceedings of the 5th International Conference on Genetic Algorithms, ICGA-93*, pages 303–309, University of Illinois at Urbana-Champaign, 17-21 July 1993. Morgan Kaufmann.

[Tac94] Walter Alden Tackett. *Recombination, Selection, and the Genetic Construction of Computer Programs*. PhD thesis, University of Southern California, Department of Electrical Engineering Systems, 1994.

[Tac95] Walter Alden Tackett. Greedy recombination and genetic search on the space of computer programs. In L. Darrell Whitley and Michael D. Vose, editors, *Foundations of Genetic Algorithms 3*, pages 271–297, Estes Park, Colorado, USA, 31 July–2 August 1994 1995. Morgan Kaufmann.

[WL96] Annie S. Wu and Robert K. Lindsay. A survey of intron research in genetics. In Hans-Michael Voigt, Werner Ebeling, Ingo Rechenberg, and Hans-Paul Schwefel, editors, *Parallel Problem Solving From Nature IV. Proceedings of the International Conference on Evolutionary Computation*, volume 1141 of *LNCS*, pages 101–110, Berlin, Germany, 22-26 September 1996. Springer-Verlag.

[ZM93] Byoung-Tak Zhang and Heinz Mühlenbein. Evolving optimal neural networks using genetic algorithms with Occam's razor. *Complex Systems*, 7:199–220, 1993.

Concepts of Inductive Genetic Programming

Nikolay I. Nikolaev[1] and Vanio Slavov[2]

[1] Department of Computer Science, American University in Bulgaria,
Blagoevgrad 2700, Bulgaria, e-mail: nikolaev@nws.aubg.bg

[2] Information Technologies Lab, New Bulgarian University,
Sofia 1113, Bulgaria, e-mail: vslavov@inf.nbu.acad.bg

Abstract. This paper presents the fundamental concepts of inductive
Genetic Programming, an evolutionary search method especially suitable
for inductive learning tasks. We review the components of the method,
and propose new approaches to some open issues such as: the sensitivity
of the operators to the topology of the genetic program trees, the coordination of the operators, and the investigation of their performance. The
genetic operators are examined by correlation and information analysis of
the fitness landscapes. The performance of inductive Genetic Programming is studied with population diversity and evolutionary dynamics
measures using hard instances for induction of regular expressions.

1 Introduction

Inductive Genetic Programming (iGP) [2], [6], [10] is a global, random search
method for evolving genetic tree-like programs. It is suitable for solving NP-
hard computational problems, many of which arise when we address inductive
tasks. Previous works point out that iGP is successful for such tasks as system
identification [3], symbolic regression [6], pattern recognition [16], scene analysis,
robot navigation, concept learning [6], [8], time-series prediction [3], etc..

Inductive Genetic Programming is a specialization of the original Genetic
Programming (GP) paradigm [6] for inductive learning [8], [12]. The reasons for
this specialized term are: 1) *inductive* learning is essentially a search problem
and GP is a versatile framework for exploration of large search spaces; 2) GP
provides *genetic* operators that can be tailored to each particular task; and 3)
GP flexibly reformulates genetic *program*-like solutions which adaptively satisfy
the constraints of the task. An advantage of iGP is that it discovers the size of
the solutions. The main iGP components are: fitness proportional selection, generational reproduction, a stochastic complexity fitness function, and the biased
application of the crossover and mutation operators [8], [12].

The evolutionary search by iGP has two aspects: navigation carried by genetic modification and selection operators, and structure determined by the fitness landscape. An important issue in iGP is the sensitivity of the navigation
operators to the topology of the genetic program trees [10], and the coordination
of their work. We review some operators [10], [12], genetic program implementations [2], [3], [6], and elaborate a novel context-preserving mutation. Recent
research suggests that a promising integration between the operators can be
achieved by biasing with the genetic program size [12]. It was found that size

biasing prevents the genetic program trees from growing or shrinking, which contributes to the progressive landscape exploration and enables efficient search.

Concerning the second structural aspect, landscape analysis is necessary for evaluating the search abilities of the navigation operators and studying their performance. Difficulties in iGP occur because the fitness landscapes are anisotropic in sense of being on non-regular underlying graphs. The iGP components may be improved through measurements of: the fitness distance correlation [4], the autocorrelation function [13], and some information characteristic [15] of their fitness landscapes. The performance investigation is another crucial issue in iGP, which is still under discussion and on which there is no consensus. We propose to observe the behavior of evolutionary search with: 1) diversity measures, such as the population diameter measured by the relevant tree-to-tree distances, and the population entropy; 2) ordinary evolutionary dynamics measures, such as the average population fitness and the best fitness. The behavior of iGP is studied in this paper with benchmark instances for induction of regular expressions [14].

Genetic program implementations and operators are discussed in section two. In section three the fitness landscapes of several mutation operators are analyzed. The performance of iGP is studied in section four. Finally, a brief discussion is made and conclusions are derived.

2 Representations and Operators for iGP

The design of an evolutionary algorithm initially faces the problem how to represent the individuals and which operators to employ. An important demand to the iGP navigation operators is to account for the implementation of the genetic programs and, eventually, for the inductive character of the task.

2.1 Genetic Programs

The set of genetic programs GP with elements g can be defined as follows:

Definition 1. Let V be a vertex set from two kinds of components: *functional nodes* F and *terminal leaves* T ($V = F \cup T$). A *genetic program* $g \in GP$ is an ordered tree $s_0 \equiv g$, in which the sons of each vertex are ordered, with properties:
- it has a distinguishing *root* vertex $\rho(s_0) = V_0$;
- its vertices are labeled $\lambda : V \to \mathbb{N}$ from left to right and $\lambda(V_i) = i$;
- any functional node has a number of children, called arity $\alpha : V \to \mathbb{N}$. A terminal leaf $\rho(s_i) = T_i$ has zero arity $\alpha(T_i) = 0$;
- the children of a vertex V_i are roots of disjoint subtrees $s_{i1}, s_{i2}, ..., s_{ik}$, where $k = \alpha(V_i)$. We specify a subtree s_i, by a tuple consisting from its root $\rho(s_i) = V_i$, and the subtrees $s_{i1}, ..., s_{ik}$ at its immediate k children: $s_i = \{(V_i, s_{i1}, s_{i2}, ..., s_{ik}) \mid k = \alpha(V_i)\}$.

This labeling of the vertices means that the subtrees below a vertex V_i are ordered from left to right as the leftmost child s_{i1} has the smallest label among the other $\lambda(s_{i1}) < \lambda(s_{i2}) < ... < \lambda(s_{ik})$. Two genetic subtrees s_i and s_j with $\rho(s_i) = V_i$ and $\rho(s_j) = V_j$, are related by *inclusion* $s_i \subset s_j$ when the arities of their roots differ by one $\alpha(V_j) - \alpha(V_i) = 1$, and such that each of the subtrees from s_i is topologically the same as exactly one subtree from s_j.

2.2 Tree Implementations

There are different possibilities for coding genetic program trees. From the implementation point of view, these are pointer-based trees, linear trees in prefix notation and linear trees in postfix notation [5]. From the logical point of view, these are parse trees [6], decision trees [8], and non-linear trees [2],[3]. Often *linear tree* implementations for the genetic programs are preferred as they allow fast processing [5]. This is of critical importance for construction of practical iGP systems. We adopt the *prefix tree* notation for the genetic programs which corresponds to the chosen labeling and could be made by preorder tree traversal.

The ways in which a tree is changed determine the sampling of its neighboring trees. An iGP operator that modifies slightly a genetic program tree should cause slight changes of its fitness. Efficient is this operator which maintains a high correlation between the fitness of the parent and this of the offspring.

The operators for linear genetic trees have to meet the following criteria [7]:
- the parent-child relationships among the genetic tree vertices remain;
- the sequence of prefix ordering, $\lambda(s_i) < \lambda(s_j)$ in g, does not change after applying a series of operators $\lambda(M(s_i)) < \lambda(M(s_j))$ in g'.

In order to achieve high correlation the operator has to sample only the closest neighbors of the selected vertex. This requires preservation of the inclusion property between the corresponding subtrees from the parent s_i and the offspring s_j trees: $s_j \subset M(s_i)$ or $s_i \subset M(s_j)$. The transformations that satisfy these three topological properties of the genetic program trees we call *context-preserving*.

2.3 Reproduction and Crossover

The most recent version of iGP uses *fitness-proportional reproduction* with stochastic universal sampling. This strategy imposes the least pressure to fit individuals during reproduction, compared to the other selection strategies.

The *crossover operator* for iGP cuts two selected programs only if the cut probability $p_c = p_k/(l-1)$ is greater than a random number p, $0 <= p <= 1$. Here, p_k is a constant, $p_k = 9$ and l is the program length. The program length is defined as the number of all nodes in a tree-like genetic program, it is the sum of functional and terminal (leaf) nodes. The cut points are randomly selected within the programs with equal probability 50% for choosing a function or a leaf node, and the subtrees after the cut points are spliced.

2.4 Context-Preserving Mutation

We propose a *context-preserving* mutation operator for local search. This mutation edits a genetic program tree subject to two restrictions: 1) maintaining the approximate topology of the genetic program tree, by keeping the representation relationships among the tree vertices; and 2) preserving of the inclusion property between the subtrees. This operator modestly transforms a tree so that only the closest neighboring vertices to the chosen mutation point are affected. The genetic program trees shrink and grow slightly, which contributes to the overall improvement of the evolutionary search behavior.

Definition 2. A genetic program tree $g \in GP$, built from vertices V_i with children subtrees $s_i = \{(V_i, s_{i1}, s_{i2}, ..., s_{ik}) \mid k = \alpha(V_i)\}$, is subject to the following *elementary mutations* $M : GP \times V \times V \rightarrow GP'$, $M(s, V_i, V_i') = s'$:

- *insert* M_I: adds a subtree $s_i' = \{(V_i', s_{i1}', T_{i2}', ..., T_{il}') \mid l = \alpha(V_i')\}$, so that the old subtree s_i becomes leftmost child of the new V_i', i.e. $s_{i1}' = s_i$;

- *delete* M_D: moves up the only subtree $s_i' = \{(F_i', s_{i1}', ..., s_{il}') \mid 1 \leq l \leq \alpha(F_i')\}$ of s_i iff $\exists F_{ij} = \rho(s_{ij})$, for some $1 \leq j \leq \alpha(V_i)$ to become root $F_i' = F_{ij}$, and all other leaf children $\forall k, ik \neq j, \rho(s_{ik}) = T_{ik}$, of the old V_i are pruned ;

- *substitute* M_S: replaces a leaf $T_i = \rho(s_i)$, by another one T_i', or a functional $F_i = \rho(s_i)$ by F_i'. If the arity $\alpha(F_i') = k$, then $s_i' = \{(F_i', s_{i1}, ..., s_{ik}) \mid k = \alpha(F_i')\}$. When $\alpha(F_i') = l$ only $l = k \pm 1$ is considered. In case $l = k + 1$ it adds a leaf $s_i' = \{(F_i', s_{i1}, ..., s_{ik}, T_{il})\}$ else in case $l = k - 1$ it cuts $s_i' = \{(F_i', s_{i1}, ..., s_{il})\}$.

The *mutation operator* in iGP is also applied with a bias proportional to the lengths of the genetic programs. As a result of biasing only trees of sufficient length are modified with probability $p_m = p_j \times (l - 1)$, where $p_j = 0.06$ [1], [12].

2.5 Tree-to-Tree Distance

The notion of distance is essential for the search algorithms as it is associated with the landscape ruggedness and with the correlation between the genetic program fitnesses. The *tree-to-tree distance* is a measure of topological similarity of trees. The distance between trees is calculated with the minimal composition of elementary mutations $M \in [M_I, M_D, M_S]$ which repeatedly convert one of them into the other [2],[7],[10]. The tree-to-tree distance here is defined with the costs of the above mutations: $\sharp(M_S(s) = s')$ is 1 if $\rho(s) \neq \rho(s')$, and 0 otherwise; $\sharp(M_I(s)) = \sharp(M_D(s)) = |s| - 1$, where $|s|$ is the number of vertices in s.

Definition 3. Distance d between two genetic program trees g and g' is the minimal number of elementary mutations M needed to produce g' from g:
$$d(g, g') = \min\{\sharp M \mid M \in [M_I, M_D, M_S], \ M(g) = g'\}$$

This tree-to-tree distance is a distance metric as it meets the formal mathematical requirements: 1) it is symmetric $d(g, g') = d(g', g)$; 2) it is nonnegative $d(g, g') = 0$ only if $g = g'$; and, 3) it obeys the triangle inequality.

We develop an original algorithm for computing the tree-to-tree distance, with respect to the above elementary mutations, that relies on preorder traversal of the trees. When two trees: g, of vertices V_i with children s_{im}, and g', of vertices V_j' with children s_{jn}', are compared it chooses the minimum among:

1) the minimum distance between all combinations of the children $s_i = \rho(V_i)$, and $s_j' = \rho(V_j')$ of roots $V_i \equiv V_j'$ plus one substitution: $\min\{d(s_i, s_j') + \sharp M_S(V_i)\}$;

2) the minimum distance between the first tree and the root children subtrees of the second tree $s_{j1}', ..., s_{jn}'$, only if mutation is possible, plus the cost of one insertion: $\forall 1 \leq l \leq \alpha(V_j'), s_{jn}', \min\{d(s_i, s_{jn}') + \sharp M_I(V_j')\}/$ here $\sharp M_I(V_j') = 1/$;

3) the minimum distance between the root children subtrees $s_{i1}, ..., s_{im}$ of the first tree against the second tree, only if mutation is possible, plus the cost of one deletion: $\forall k, 1 \leq k \leq \alpha(V_i), \min\{d(s_i, V_j') + \sharp M_D(V_i)\}/$ here $\sharp M_D(V_i) = 1/$.

The limitation if mutation is possible requires checking whether the descendants of the roots differ by one.

3 Fitness Landscapes in iGP

The evolutionary search difficulties strongly depend on the structure of the fitness landscape. Fitness landscape is the set of genetic programs, a fitness function which assigns a value to each genetic program, and a neighborhood relation determined by the navigation operator. It is widely assumed that a rugged landscape is hard to search, while a smooth landscape is usually easy to search. As the fitness function determines the character of the landscape, we have to choose carefully fitness functions that produce relatively smooth profiles.

3.1 The Fitness Function

The stochastic complexity (or minimum description length) principle [11] is suitable for GP fitness functions [3], [9], [16]. This principle recommends that most significant is this genetic program that possesses low syntactic complexity, i.e. has a small size, and high predictive accuracy, i.e. covers accurately most of the examples. Such a *stochastic complexity fitness function* should be employed in iGP in order to identify parsimonious and useful genetic tree topologies [12].

3.2 Global and Local Landscapes

The structure of the fitness landscape provides valuable information for describing the efficacy of a navigation operator. A fitness landscape analysis may help to tune the parameters of an operator for improving the search, and to determine the proper one among several alternatives. We distinguish between local and global characteristics of a fitness landscape. The *global landscape* characteristics depend on the correspondence between the fitness and the distance to some known global optima. This is evaluated with the fitness distance correlation [4]. The *local landscape* characteristics depend on the relation between the fitness and the operator. This is estimated by the autocorrelation function [13].

Experiments for inducing regular expressions [14] were performed with three operators: *context-preserving mutation* (CPM), *hierarchical variable length mutation* (HVLM) [10], and *replacement mutation* (RM) [12]. HVLM inserts a subtree before a chosen vertex, deletes a node so that its largest subtree is promoted to replace it, and substitutes a chosen node by another one with the same arity. RM replaces probabilistically each successively selected vertex in the tree by a different randomly chosen functional node or a randomly chosen terminal leaf.

The *global characteristics* of three fitness landscapes were estimated taking random samplings of 20000 arbitrary genetic programs. The tree-to-tree distances $d(g, g^*)$, measured with respect to each of the above mutation operators, from the current tree g to the globally optimal tree g^* were registered and used to calculate the fitness distance correlation φ as follows [4]:

$$\varphi \overset{def}{=} \frac{\langle f(g)d(g,g^*)\rangle - \langle f\rangle\langle d\rangle}{\langle fd\rangle - \langle f\rangle\langle d\rangle}$$

The fitness distance correlations were: 1) from the CPM landscape, in the interval $[0.6946 \div 0.8837]$; 2) those from the HVLM landscape, in $[0.6371 \div 0.8444]$;

and those from the RM landscape, in $[0.5781 \div 0.6124]$. These values indicate that the three fitness landscapes are straightforward according to the original definition in [4]. Despite this, the CPM operator has a higher fitness distance correlation, hence its landscape is globally smoother than the other two.

The *local characteristics* of three fitness landscapes were also identified. On each landscape we sampled randomly 20000 pairs of genetic programs (g_1, g_2) at a particular tree-to-tree distance $d(g_1, g_2) = n$ with respect to the corresponding operator, using their fitnesses f in the following autocorrelation function ρ [13]:

$$\rho(n) \overset{def}{=} \frac{\langle f(g_1)f(g_2)\rangle_{d(g_1,g_2)=n} - \langle f \rangle^2}{\langle f^2 \rangle - \langle f \rangle^2}$$

The experiments were performed 10 times each with a chosen tree-to-tree distance from $d = 1$ to $d = 10$ and the results averaged. The plots on Figure 1 show that the autocorrelation of CPM is higher than that of the other two operators. The fitness of a genetic program sampled by CPM seems to be relatively similar to the fitnesses of its relatively distant programs, hence CPM searches on a locally smoother fitness landscape than the other two operators. We hypothesize that CPM mitigates the search difficulties compared to HVLM and RM, as it transforms the genetic program trees more flexibly. The fitness evaluations were made with boolean strings from Tomita's set 7 [14].

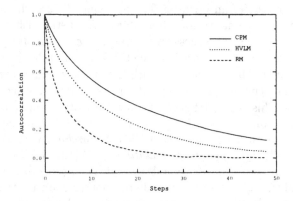

Figure 1. Autocorrelations of CPM, HVLM and RM each
averaged over 10 different distances on their landscapes

The correlation landscape characteristics are only a necessary conditon for the usefulness of the operators and the fitness function. However, they do not guarantee a good performance of the evolutionary algorithm. From a local perspective, a landscape with a single hill could have a high autocorrelation but the evolutionary algorithm may climb slowly making a long path on its slopes before reaching the global optima. From a global perspective, a landscape may be featured by a good fitness distance correlation but if the global optima is surrounded by a lot of local optima it will be arduous for the algorithm to search.

3.3 Information Characteristics

The fitness landscape may be examined from the perspective of its *information content*. Recently a measure which evaluates the informativeness of the landscape ruggedness was formulated [15]. The information content of a sequence of fitnesses $F = \{f_0, f_1, ..., f_n\}$, obtained by a random walk on a landscape, is defined as the entropy H of the profile structure of the landscape represented as a string $Q(\varepsilon) = \{q_0, q_1, ..., q_n\}$. Each string element $q \in \{-1, 0, 1\}$ and $q_i = \sigma_F(i, \varepsilon)$ is a particular substructure component determined by the function [15]:

$$\sigma_F(i, \varepsilon) = \begin{cases} -1, & \text{if } f_{i+1} - f_i < -\varepsilon \\ 0, & \text{if } | f_{i+1} - f_i | \le \varepsilon \\ 1, & \text{if } f_{i+1} - f_i > \varepsilon \end{cases}$$

The entropy is calculated with a real number $\varepsilon \in length[0, \max(f) - \min(f)]$:

$$H(\varepsilon) = - \sum_{p \neq q} P_{\{pq\}} \log P_{\{pq\}}$$

where P are the probabilities of substrings $\{pq\}$ of elements $\{-1, 0, 1\}$.

The information content reveals whether the peaks remain when we observe the landscape from more and more distant points of view, which indicates whether these are high peaks of a rugged landscape (Figure 2).

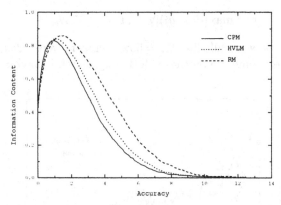

Figure 2. Information content curves, each calculated with 20000 points
of the fitness landscapes of CPM, HVLM and RM

Figure 2 make us certain that CPM has a smoother landscape than the other mutations. With respect to the intervals of the information accuracy, the explanation is: 1) the fact that the curves start from the common point 0.4 means that the three operators have landscapes with approximately the same diversity of peaks; 2) the curve raisings in [0,1] demonstrate that all the operators climb or descend unevenly on their landscapes which are rugged; 3) within [1,10] one realizes that the number and height of the unessential peaks on the landscape of CPM is smaller than these on the landscapes of HVLM and RM, or the CPM landscape is smoother; and, 4) the shared termination of the curves at 12.5 denotes that all landscapes have the same number of essential peaks.

4 Performance of iGP

The performance of iGP can be evaluated by estimates of the population diversity, and estimates of the evolutionary dynamics. We present empirical results obtained with a classical NP-hard problem, the induction of regular expressions. The experiments conducted using well known benchmark sets of strings [14].

An evolutionary algorithm like iGP may visit potentially every landscape region if a high *population diversity* is maintained. When the variety of genetic programs in the population decreases, the evolution exploits only a small region on the landscape and the search stagnates. The diversity depends on two factors: 1) it is due to the selection pressure; and 2) it is sensitive to the sampling performed by the navigating crossover and mutation operators. The last factor may be addressed by coordinating of the navigation operators with appropriate biasing. Biasing in iGP is called the policy for application of the operators. The current versions of iGP use biased crossover, which cuts large programs and splices small programs in case of fitness functions with minimizing effect, and biased mutation that transforms only large size programs [12].

We investigate the ability of iGP to maintain high diversity of genetic programs with two estimates: the population diameter and the structural population entropy. *Population diameter* π is the maximum tree-to-tree distance among the elite individuals that have best fitness values:

$$\pi = \max\{d(g_i, g_j)|\forall i \forall j, 1 \leq i, j \leq N_{el}\}$$

The distances between all elite $N_{el} = 50\%$ members were calculated because they have strongest influence on the evolutionary search behavior.

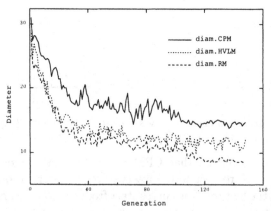

Figure 3.Population diameter, from runs with 50 programs using Tomita's set 7

The averaged curves from 10 runs in Figure 3 are a testimony that the CPM operator supports higher variety, which is one of the reasons for the iGP using CPM to outperform the iGP implementations with HVLM and RM. This is an empirical proof that iGP evolves better solutions equipped with CPM because then it continuously promotes more dissimilar individual genetic programs.

The *structural population entropy* of the iGP system we calculate as sum of the entropies of all genes divided by the length of the largest genetic program. The contribution of a single gene is the entropy of its relative frequencies from the corresponding column in the prearrangement of the genetic programs one by one in a population table. Each column of this population table is of size equal to the population size, and contains the alleles from the corresponding loci, which are functional nodes or terminal leaves. The structural population entropy is the sum of entropies of all genes divided by the size of the longest genetic program.

The tendency in evolutionary search should be to order and a decreasing of the randomness. The averaged entropy plot of CPM in Figure 4 declines variably which means that the population self-organizes its emerging structure. It seems that using the HVLM or RM the system is not able to keep in the population genetic program structures that carry useful information.

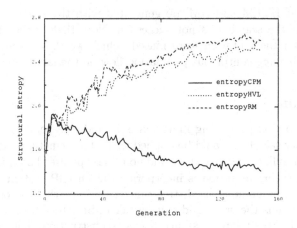

Figure 4. Population Entropy, from runs with 50 programs using Tomita's set 7

The question of how to measure the *evolutionary dynamics* of the genetic algorithms is still open for research. An evaluation of the dynamics by ordinary estimates of the average and best population fitnesses is suitable for characterizing the behavior of search algorithms when solving inductive tasks as these reflect the degree of improving the learning accuracy. That is why, in a methodology for analysis of the performance of iGP we may include diagrams of the changes of average and best fitnesses. These measures, however, are not sufficient to claim that if an iGP system produces decreasing performance curves, in case of minimizing fitness function, it will always search efficiently.

The plots given below (Figure 5) are recorded during runs again with the Tomita's set 7, but they are typical and coincide with the curves found with the remaining six data sets of strings [14]. What we see on Figure 5 is that iGP evolves increasingly fit genetic programs, and it progressively improves its learning accuracy. The correct regular expression was identified always using CPM in cooperation with crossover and stochastic complexity fitness.

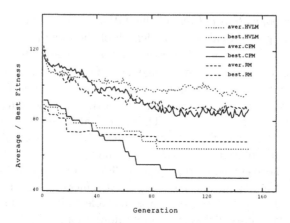

Figure 5. Evolutionary dynamics, from runs with 50 programs using Tomita's set 7

The curves of HVLM and RM are generated with the same iGP settings, and despite this the search was not successful. Note that the applications of the crossover and the mutation were biased, which we think is an important condition for achieving continuous population flow on the landscape.

5 Discussion

The presented context-preserving mutation is a kind of a general operator for local evolutionary search. It could be applied with other genetic program implementations such as linear postfix trees, and non-linear pointer-based trees. When the context-preserving mutation is incorporated in an iGP system with other, postfix or non-linear pointer-based, tree representations, the proposed tree-to-tree distance remains the same and can be taken directly. Only the algorithm for computing the tree-to-tree distance needs reconstruction. The requirement is to adjust in advance the measure with respect to the chosen mutation.

6 Conclusion

This paper sketched the development of Genetic Programming search systems suitable for inductive learning tasks. The stress was on two aspects which we find very essential for iGP: first, the operators should be carefully crafted and coordinated in order to achieve efficient search; second, the performance measures should allow the observation of the diversity and the evolutionary dynamics. Concerning the first aspect, we demonstrated that the careful design of an operator, here the mutation, may increase the search power of iGP. It was shown also that size proportional biasing of the the genetic operators and the fitness function imply successful evolutionary search.

Concerning the second aspect, the population diameter and population entropy were considered reliable diversity estimates. It is interesting to think about elaboration of other evolutionary dynamics measures for iGP. The question for further research is: can we measure the dynamics of evolutionary algorithms with estimates formulated in analogy with the physics of complex systems ?

References

1. D. Goldberg, B. Korb and K. Deb, 'Messy Genetic Algorithms: Motivation, Analysis and First Results', *Complex Systems*, 3:493-530, 1989.

2. H. Iba and T. Sato, 'Meta-level Strategy for Genetic Algorithms based on Structured Representations', In: *Proc. Second Pacific Rim Int. Conf. on Artificial Intelligence*, pp.548-554, 1992.

3. H. Iba, H. de Garis and T. Sato, 'Genetic Programming using a Minimum Description Length Principle', In: *Advances in Genetic Programming*, K.Kinnear Jr.(ed.), The MIT Press, 265-284, 1994.

4. T. C. Jones and S. Forrest, 'Fitness Distance Correlation as a Measure of Search Difficulty for Genetic Algorithms', In: *Proc. Sixth Int. Conference on Genetic Algorithms*, L.Eshelman (ed.), 184-192, 1995.

5. M.J. Keith and M.C. Martin, 'Genetic Programming in C++: Implementation Issues', In: K.E.Kinnear Jr.(ed.),*Advances in Genetic Programming*, The MIT Press, Cambridge, MA, 285-310, 1994.

6. J. R. Koza, *Genetic Programming: On the Programming of Computers by Means of Natural Selection*, The MIT Press, Cambridge, MA, 1992.

7. S.Y. Lu, 'A Tree-to-Tree Distance and its Application to Cluster Analysis', *IEEE Trans. on Pattern Analysis and Machine Intelligence*, 1(1):219-224, 1979.

8. N. Nikolaev and V. Slavov. 'Inductive Genetic Programming with Decision Trees', In: M.van Someren and G. Widmer (eds.), *Machine Learning: ECML-97, Ninth European Conf. on Machine Learning*, LNAI-1224, Springer, Berlin, 183-190, 1997.

9. P. Nordin and W. Banzhaf, 'Complexity Compression and Evolution', In: L. Eshelman (ed.), *Proc. Sixth Int. Conf. on Genetic Algorithms, ICGA-95*, Morgan Kaufmann, CA, 310-317, 1995.

10. U.-M. O'Reilly, *An Analysis of Genetic Programming*, PhD Dissertation, Carleton University, Ottawa, Canada, 1995.

11. J. Rissanen, *Stochastic Complexity in Statistical Inquiry*. World Scientific Publishing, Singapore, 1989.

12. V. Slavov and N. Nikolaev, 'Inductive Genetic Programming and the Superposition of the Fitness Landscape', In: T. Bäck (ed.), *Proc. Seventh Int. Conf. on Genetic Algorithms, ICGA-97*, Morgan Kaufmann, CA, 97-104, 1997.

13. P. Stadler, 'Towards a Theory of Landscapes', In: *Complex Systems and Binary Networks*, R.Lopéz-Peña et al.(eds.), Springer-Verlag, Berlin, 77-163, 1995.

14. M. Tomita, 'Dynamic construction of finite-state automata from examples using hill climbing', In: *Proc. Fourth Annual Conf. of the Cognitive Science Society*, Ann Arbor, MI, 105-108.

15. V. Vassilev, 'An Information Measure of Landscapes', In: T. Bäck (ed.), *Proc. Seventh Int. Conf. on Genetic Algorithms, ICGA-97*, 49-56, 1997.

16. B.-T. Zhang and H. Muhlenbein, 'Balancing Accuracy and Parsimony in Genetic Programming', *Evolutionary Computation*, 3:1, 17-38, 1995.

Immediate Transfer of Global Improvements to All Individuals in a Population Compared to Automatically Defined Functions for the EVEN-5,6-PARITY Problems

Ricardo Aler

Universidad Carlos III de Madrid
Butarque 15
28911 Leganés (Madrid), España
aler@inf.uc3m.es
http://grial.uc3m.es/~aler

Abstract. Koza has shown how automatically defined functions (ADFs) can reduce computational effort in the GP paradigm. In Koza's ADF, as well as in standard GP, an improvement in a part of a program (an ADF or a main body) can only be transferred via crossover. In this article, we consider whether it is a good idea to transfer immediately improvements found by a single individual to the whole population. A system that implements this idea has been proposed and tested for the EVEN-5-PARITY and EVEN-6-PARITY problems. Results are very encouraging: computational effort is reduced (compared to Koza's ADFs) and the system seems to be less prone to early stagnation. Finally, our work suggests further research where less extreme approaches to our idea could be tested.

1 Introduction

In [4], Koza showed how "automatically defined functions enable genetic programming to solve a variety of problems in a way that can be interpreted as a decomposition of a problem into subproblems, a solving of the subproblems, and an assembly of the solutions to the subproblems into a solution to the overall problem". Also, he showed that "For a variety of problems, genetic programming requires less computational effort to solve a problem with automatically defined functions than without them, provided the difficulty of the problem is above a certain relatively low problem-specific breakeven point for computational effort".

With Koza's ADFs, each individual consists of both the main body of the program and of all its subroutines. An improvement in a subroutine (or a main body) of an individual can only be transferred to another individual via crossover between both individuals. Intuitively, it would seem that if an individual can immediately use improvements obtained anywhere else in the population, then the rate of discovery would be higher. On the other hand, it might happen that an individual cannot use another's individual discovery because their structures are just too different. In that case, something could be an improvement for an individual but a hindrance to another and therefore, former good individuals would become instantly bad individuals. What would be the net effect of both tendencies is unclear. Thus, the aim of this article is to start exploring empirically this matter: what would happen if improvements[1] in a subroutine (or a program's main body) would be transferred immediately to all individuals of the population?. Explaining how this idea has been implemented is the purpose of the next section.

2 Implementation

Let us suppose that the architecture of our individuals consists of just two parts: a main body and one ADF (ADF0). As it is shown in Figure 1, our implementation divides the population of individuals into two separated populations: one for program's main bodies (called main population) and the other one for ADFs (named ADF population). Both populations will evolve independently. Each population will supply the best individual obtained so far to the other population so that individuals of the other population can be evaluated. More specifically, the main population will supply the best main body obtained so far to the ADF population. Likewise, the ADF population will supply the best ADF0 obtained so far to the main population.

In order to evaluate a member (a main body) of the main population, it will be coupled with the best ADF0 supplied from the ADF population, a whole individual will be built and the fitness obtained by that individual will be assigned to the main body being evaluated. Similarly, in order to evaluate an ADF in the ADF population, the ADF will be coupled with the best main body supplied by the main population and the resulting individual will be evaluated. Of course, it is impossible to evaluate an individual in a population until a best individual has been obtained in the other population. But this is also true for individuals in the other population. As the process must start somewhere, at the beginning of the run a randomly chosen individual from each population is designated as the best of that population.

---

[1] The reader should be aware that by *improvements* we mean *global* improvements, that is, improvements that lead to a change in the best of population

Therefore, all individuals in the main population will be coupled and evaluated with the same ADF (the best one obtained so far). And vice versa. Thus, an improvement found by the main population will be immediately transferred to all ADFs in the ADF population (and vice versa). An immediate advantage of this idea over Koza's ADF is that each population can run in a separate machine, being the interaction between populations (and therefore, network communication) very low. Although figure 1 displays only two populations, many more populations could be used in problems requiring more ADFs. Our implementation doesn't take advantage of parallelism, though, so populations are evaluated sequentially: if we let a n population system run for 150 generations, P_0 will be run at generation 0, P_1 will be run at generation 1 and so on. P_0 will be

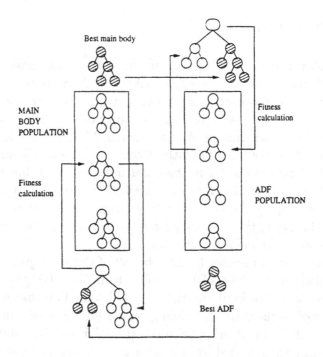

Fig. 1. Inter-relations between the main body population and the ADF population.

run again at generation n, P_1 at generation $n+1$ and so forth. Therefore, each P_i will run for $150/n$ generations, interleaved with the rest of the populations. Table 1 shows the algorithm we have used for this article.

1. For each i, create population P_i and choose randomly a best-of-run B_i.
2. Do until the number of generations is exhausted or success.
   (a) run GP on P_0 for 1 generation and update B_0. Stop if success.
   (b) run GP on P_1 for 1 generation and update B_1. Stop if success.
   (c) ...
   (d) run GP on P_{n-1} for 1 generation and update B_{n-1}. Stop if success.

Table 1. Basic *Iadf* algorithm.

Next section shows some experimental results obtained by our system for the EVEN-5-PARITY and EVEN-6-PARITY problems.

3   Experimental Results

The system shown in figure 1 has been tested with the EVEN-5-PARITY and EVEN-6-PARITY problems, described in [4]. Table 3 shows the tableau for the EVEN-5-PARITY problem problem with ADFs (the EVEN-6-PARITY problem is similar).

This tableau is similar to Koza's but for M and G^2. Koza's M is 16000 whereas we use a much smaller population size of 200 for EVEN-5-PARITY and 400 for EVEN-6-PARITY. Besides, as we wanted to explore the behaviour of the system for long runs, our G has been extended to 150 (being Koza's $G = 51$). As we use different parameters, we performed a series of experiments for Koza's ADFs as well, so that it can be compared to our system. From now on, Koza's ADF results will be referred to as *Kadf* and our system results as *Iadf* ("Independent ADFs"). As Koza states in [4], a good way to determine how well an adaptive system performs (for a given problem and chosen parameters) is to obtain the computational effort (E) for that problem. Computational effort and related data for both *Kadf* and *Iadf* are shown in table 3. Graphs displaying computational effort per generation are shown in figures 2 and 3. Also, figures 4 and 5 show the cumulative probabilities of solving EVEN-5-PARITY and EVEN-6-PARITY respectively.

---

[2] M is the size of the population and G is the number of generations

Objective	Find a program that produces the value for the Boolean even 5-parity function as its output when given the values of the tree independent Boolean variables as its input.
Architecture	One result-producing branch and two two-argument functions-defining branches, with ADF1 hierarchically referring to ADF0.
Parameters	Branch typing.
Terminal set for the result-producing branch:	D0, D1, D2, D3, D4
Function set for the result-producing branch:	ADF0, ADF1, AND, OR, NAND, and NOR
Terminal set for the function-defining branch ADF0:	ARG0 and ARG1
Function set for the function-defining branch ADF0:	AND, OR, NAND, and NOR.
Terminal set for the function-defining branch ADF1:	ARG0 and ARG1
Function set for the function-defining branch ADF1:	AND, OR, NAND, NOR, and ADF0 (hierarchical reference to ADF0 by ADF1).
Fitness cases	All $2^5 = 32$ combinations of the five Boolean arguments D0, D1, D2, D3, D4.
Raw fitness	The number of fitness cases for which the value returned by the program equals the correct value of the even-5-parity function.
Standardized fitness	The standardized fitness of a program is the sum, over the 32 fitness cases, of the Hamming distance (error) between the value returned by the program and the correct value of the Boolean even-5-parity function.
Hits	Same as raw fitness.
Wrapper	None.
Parameters	$M = 200$, $G = 150$
Success predicate	A program scores the maximum number of hits

Table 2. Tableau with ADFs for the even-5-parity problem.

	EVEN-5-PARITY		EVEN-6-PARITY	
	Kadf	*Iadf*	*Kadf*	*Iadf*
Number of experiments	194	200	111	84
Population size	200		400	
Effort $= min_{i=0..149}(I(M, i, 0.99))$	408000	359600	1550000	627200
Best generation $i*$	33	30	30	48
E_{Kadf}/E_{Iadf}	1.134594		2.471301	

Table 3. Computational effort results (and related data) for *Kadf* and *Iadf*.

4 Discussion

It turns out that *Iadf* performs slightly better than *Kadf* (see table 3) for the EVEN-5-PARITY problem (being the effort ratio $E = 1.134594$) and much better for the more complex EVEN-6-PARITY problem ($E = 2.471301$). However, that is the effort that the system would have spent had we chosen $G = i*$. But $i*$ is not a datum we can know a priori. Had we started our runs without this knowledge, we could have chosen any other G and spent a different computational effort I. In order to have a better picture of what happens for different values of G, graphs displaying computational effort are shown in figures 2 and 3. Also, figures 4 and 5 show the cumulative probabilities of solving EVEN-5-PARITY and EVEN-6-PARITY respectively. Results in these graphs can be easily summarized:

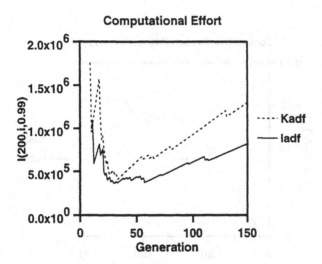

Fig. 2. Computational effort (I) for both *Kadf* and *Iadf*, given that EVEN-5-PARITY should be solved by generation i with probability z = 0.99

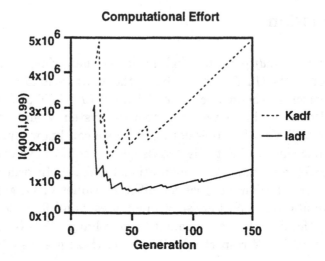

Fig. 3. Computational effort (I) for both *Kadf* and *Iadf*, given that EVEN-6-PARITY should be solved by generation i with probability z = 0.99

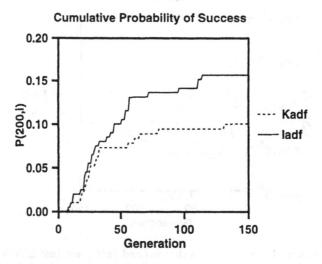

Fig. 4. Cumulative probability of solving EVEN-5-PARITY by generation i with M=200 for both *Kadf* and *Iadf*

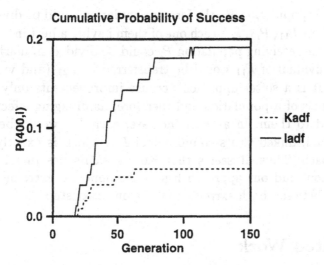

Fig. 5. Cumulative probability of solving EVEN-6-PARITY by generation i for both *Kadf* and *Iadf*

- *Iadf* has a smaller computational effort than *Kadf* for all generations and specially for late generations (see figure 2). This fact is even more noticeable for the more difficult problem (EVEN-6-PARITY) (see figure 3).
- *Iadf* manages to keep a steady rate of improvement (in terms of cumulative probability of success) for longer than *Kadf* (see figure 4). *Kadf*'s rate diminishes by generation 30 whilst *Iadf* continues improving at a good pace for much longer. Again, this is even more noticeable in the EVEN-6-PARITY problem (see figure 5).

5 Future Work

In this article we have studied the effects of transferring improvements immediately to a whole population. Our expectations were that this might be useful in terms of computational effort, and it happened to be case for the EVEN-5,6-PARITY problems. However, in other problems it might happen that an improvement for an individual A could not always be accepted as such by another individual B and in the worst case, it will hinder that individual. Many other individuals could be hindered in such a way, being their evolutionary pathways lost as a result.

In that case, a softer way of transferring improvements might be useful. Instead of transferring the best individual of a whole population P_1 to another whole population P_2, the former population could be divided into subpopulations $P_{11}, P_{12}, \ldots$, each one of them having a best individual of their own. The receiving population P_2 could be divided similarly. Then, the best individual of P_{11} would be transferred to P_{21} (and vice versa) and so on. It is a softer approach because improvements only affect a few individuals of a population and therefore, dammaging effects would be restricted to them. In an extreme case, each P_{1i} would be exactly *one* individual, linked to its co-individual P_{2i}, which is exactly Koza's ADF approach. Thus, it seems that Koza's ADFs lies in an extreme of a gradation, and our approach lies in the opposite extreme. Testing approaches between both extremes will be our next step.

6 Related Work

Iadf originated in an idea that emerged from previous work done by the author. In [2], it was indirectly shown how fixing part of a program and letting the rest evolve, could be an interesting way for a programmer to introduce background knowledge into GP and to reduce the search space. This article is an offshoot of that idea, although it is an evolving population best individual (instead of the programmer) which fixes part of the program for the rest of the evolving populations.

The system we have studied in this paper can be considered as an extreme case of co-evolution [3], albeit a strange one, because interaction between populations happens only through the best individual of each population. Co-evolution of a main program and several independent ADF populations has already been dealt with in [1]. Both approaches differ in perspective, though: we are more interested in the simultaneous transfer of information from one population to all individuals in the other populations than in studying general ADF co-evolution. In their approach, in order to evaluate a main program, ADF individuals are selected from the ADF sub-populations. They tested several selection policies, being "the best individual" policy very similar to our own approach. However, in their work this policy doesn't fare well compared to GP+ADF, which is the opposite of the results obtained in our paper. Other differences are that we use a generational model for all evolving populations instead of a steady-state model and that we favor program-level fitness evaluation instead of evaluating directly the individuals in the ADF sub-populations. Finally, our results are in terms of computational effort to solve the problem rather than of average results per generation, as in their case.

7 Conclusions

This paper started by posing the question of whether it would be useful that improvements in a part of an individual (a subroutine, for instance) would be transferred to all members of the population as soon as they were found. We then proposed a system to test this idea and utilized it for the EVEN-5,6-PARITY problems. A comparison of our results with Koza's ADF applied to the same problem shows that performance (in terms of minimum computational effort) is better. Thus, there seems to be an advantage by immediately transferring improvements to all individuals, at least in this case.

Our approach is another way to parallelize GP, with the advantage that communication between populations happens at a very small rate: all the information populations need to exchange is the best individual obtained so far, which changes rather slowly. This kind of parallelism would be useful for problems requiring many different ADFs.

Our system shows a curious effect: the cumulative probability of success keeps increasing at a good rate for longer than $Kadf$. That is, it doesn't seem to stagnate as soon as GP (or GP+ADF) does.

The approach scales well, obtaining better results for the more complex problem than for the simpler one.

It has been shown that our approach and Koza's ADFs lie in opposite extremes of a gradation. Thus, our work suggests that testing approaches between both extremes might be an interesting line of research.

Finally, we are well aware that in order to draw general conclusions beyond the problems studied in this article, many more experiments must be carried out for different problems and different parameters.

References

1. Manu Ahluwalia, Larry Bell, and Terence C. Fogarty. Co-evolving functions in genetic programming: A comparison in ADF selection strategies. In John R. Koza, Kalyanmoy Deb, Marco Dorigo, David B. Fogel, Max Garzon, Hitoshi Iba, and Rick L. Riolo, editors, *Genetic Programming 1997: Proceedings of the Second Annual Conference*, pages 3–8, Stanford University, CA, USA, 13-16 July 1997. Morgan Kaufmann.
2. Ricardo Aler, Daniel Borrajo, and Pedro Isasi. Evolving heuristics for planning. In *Proceedings of the Seventh Annual Conference on Evolutionary Programming*, Lecture Notes in Artificial Intelligence, San Diego, CA, March 1998. Springer-Verlag.

3. W. Daniel Hillis. Co-evolving parasites improve simulated evolution as an optimization procedure. In Christopher G. Langton, Charles Taylor, J. Doyne Farmer, and Steen Rasmussen, editors, *Artificial Life II*, volume X of *Sante Fe Institute Studies in the Sciences of Complexity*, pages 313–324. Addison-Wesley, Santa Fe Institute, New Mexico, USA, February 1990 1992.
4. J.R. Koza. *Genetic Programming II*. The MIT Press, 1994.

Non-destructive Depth-Dependent Crossover for Genetic Programming

Takuya ITO[1], Hitoshi IBA[2] and Satoshi SATO[1]

[1] School of Information Science, Japan Advanced Institute of Science and Technology, 1-1 Asahidai, Tatsunokuchi, Nomi, Ishikawa, 923-12 Japan
[2] Machine Inference Section, Electrotechnical Laboratory, 1-1-4 Umezono, Tsukuba, Ibaragi, 305 Japan

Abstract. In our previous paper [5], a depth-dependent crossover was proposed for GP. The purpose was to solve the difficulty of the blind application of the normal crossover, i.e., building blocks are broken unexpectedly. In the depth-dependent crossover, the depth selection ratio was varied according to the depth of a node. However, the depth-dependent crossover did not work very effectively as generated programs became larger. To overcome this, we introduce a non-destructive depth-dependent crossover, in which each offspring is kept only if its fitness is better than that of its parent. We compare GP performance with the depth-dependent crossover and that with the non-destructive depth-dependent crossover to show the effectiveness of our approach. Our experimental results clarify that the non-destructive depth-dependent crossover produces smaller programs than the depth-dependent crossover.

1 Introduction

The normal (canonical) crossover of GP chooses a node randomly regardless of its depth[1][7, pp. 599], i.e., the node selection ratio is chosen to be uniform. The node selection ratio is the probability of selecting a node as a crossover point. This means that the normal crossover has a sort of blindness character which may not work well for the effective recombination or for the accumulation of "building blocks". A "building block" is an effective small program which contributes to generating a solution [2]. To solve this difficulty, we have proposed a "depth-dependent crossover", by which the depth selection ratio for a crossover is higher for a node closer to the root node [5]. There have been several related studies which attempted to improve crossover operation and accumulate building blocks[5]. However, we have proposed the depth-dependent crossover to enhance the influence of the crossover operation for accumulating building blocks via the depth selection ratio. The depth-dependent crossover contributed to the reduction of the computational time for the evolution on boolean concept formation problems. However, there seemed to be a difficulty that the depth-dependent crossover generated very large programs. Large program sizes mean

---

[1] The depth of a node v in a tree is defined to be the distance from the root node to v [1, p. 53]

that the time necessary to measure their fitness often dominate total processing time. Large programs require huge memories to evolve programs.

There have been several related studies which tackled the problem of GP program size. Soule has considered that shorter programs tend to show better generalization performance than longer programs. Thus he has studied the selective pressure by penalizing longer programs [10]. This method appeared to be effective in bounding the programs' size. Kinnear added the inverse size of generated programs to the fitness measure [6]. However, its fitness measure had the size factor (sf) by which the size of the program was multiplied for addition. We cannot know in advance the exact value of the factor. Iba has introduced a method for controlling program (tree) growth, which used an MDL (Minimum Description Length) principle to define GP fitness functions [3]. However, MDL-based fitness functions could not be applied to every kind of problem to be solved by GP. Trees had to have the two characteristics, i.e., "size-based performance" (the more the tree grows, the better its performance) and "decomposition" (the fitness of a substructure is well-defined itself). We adopt the "non-destructive crossover (NDC)" to reduce the program size. By this method, each offspring, i.e., a new tree resulting from the crossover operation, is kept only if its fitness is better than that of its parent [11]. The NDC works purely syntactically, i.e., it requires neither semantic heuristics nor problem-dependent parameters.

This paper is organized as follows. Section 2 explains a mechanism of the depth-dependent crossover. Section 3 described the non-destructive depth-dependent crossover. Section 4 shows several experimental results in several tasks and compares the performance of the depth-dependent crossover and that of the non-destructive depth-dependent crossover. Discussion is given in section 5, followed by some conclusion and future work in section 6.

2   Depth-Dependent Crossover

We have introduced the depth-dependent crossover, in which shallow nodes are more often chosen as the crossover points (Fig. 1).

When applying the crossover operator in GP, it has an influence on the resultant structure in two ways. First, the operator swaps larger parts of subtrees for shallower nodes. This leads to the propagation of larger parts of useful subtrees to the entire population. Second, the operator encapsulates a larger part of a tree, so that substructures for shallower nodes are protected from the destructive crossover. As a result, the influence of crossover is greater for shallower nodes than for deeper nodes. Therefore, by using the depth-dependent crossover, building blocks (i.e., effective partial tree structures for generating a solution) are expected to be encapsulated against the destructive crossover, and to be accumulated.

The algorithm of the depth-dependent crossover is described below:

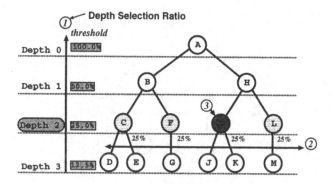

Fig. 1. Depth Selection Ratio by Depth-Dependent Crossover

STEP1. Given a tree, determine the depth d for applying the depth-dependent crossover.

STEP2. Select randomly a node of which depth is equal to d in STEP1.

STEP3. Apply the canonical crossover for the nodes chosen in STEP2 (see Fig. 1).

The depth selection ratio is derived by using the following equations:

$$\begin{cases} threshold_i = 1/2^i, & \text{if } i = depth. \\ threshold_i = threshold_{i+1} \times 2, & \text{otherwise } i = 0, 1, \cdots, depth - 1 \end{cases} \quad (1)$$

where $threshold_i$ is the depth selection ratio at the ith depth, and $depth$ is the depth of a tree. $threshold_i$ is an accumulated value from the deepest node. The above equations represent that the depth selection ratio is set to be 1.0 for a root node. The depth selection ratio is half of its parent node's ratio (see Fig. 1). When determining the depth d for the depth-dependent crossover (i.e., STEP1), we pick up a number i between 0 and the maximum depth in proportion to the $threshold_i$ value. This process is similar to the roulette wheel selection used in GA.

In Fig. 1, the depth selection ratio of a root node is 50.0% by the above definition. Note that the threshold value of a root node in the figure are those accumulated from the deepest node. On the other hand, it is 7.7% for the normal crossover (see Fig. 2).

The above depth-dependent crossover has to be improved in terms of the uniformity because it selects a node randomly among those nodes that are of the dth depth (STEP2). For instance, node C, F, I and L have the same probability by the depth-dependent crossover, regardless of their subtree size (Fig. 1). When node C is selected as a crossover point, the size of the swapped subtree is three. On the contrary, when the node L is selected as crossover point, the size is two. Therefore, node C has a greater influence on the crossover than node L. However,

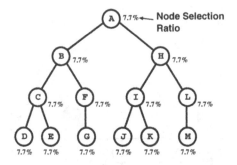

Fig. 2. Node Selection Ratio by Normal Crossover

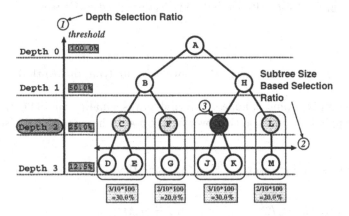

Fig. 3. Depth Selection Ratio and Subtree Size Based Selection Ratio by a Revised Version of Depth-Dependent Crossover

the depth-dependent crossover selects nodes regardless of these subtree sizes. This may decrease the influence of the crossover. Therefore, we have introduced a revised version of the depth-dependent crossover by modifying STEP2 as follows:

STEP2'. Select a node in proportion to its subtree size. The depth of the root node of the subtree is equal to d in STEP1 (see Fig. 3).

We call subtree size based selection ratio.

3   Non-Destructive Depth-Dependent Crossover

We observed that the depth-dependent crossover as well as the revised depth-dependent crossover produced a much larger program than the normal crossover [5]. To solve this difficulty, we have introduced the "non-destructive crossover" based on [11]. The NDC is a crossover operation in which each offspring is kept only if its fitness is better than that of its parents. One benefit of using the NDC is that the program growth is reduced to only the growth necessary for improving program fitness[11, p. 314]. Thus, we use the non-destructive crossover to reduce the problem growth on the depth-dependent crossover. We call this type of crossover as the non-destructive depth-dependent crossover, and the original depth-dependent crossover as the destructive depth-dependent crossover. Therefore algorithm of the non-destructive depth-dependent crossover is given by adding the following procedure to the algorithm of the original depth-dependent crossover described in section 2.

STEP4. Each offspring is kept only if its fitness is better than that of its parent.

4 Experimental Results

We have investigated the effectiveness of the non-destructive crossover and the destructive depth-dependent crossover for several problems. This section shows the experimental results in Boolean concept formation problems and ANT problem. The task of the 11MX (11-multiplexor) problem is to decode an address encoded in binary and to return the binary data value of the register at that address. An 11-multiplexor has 3 binary-valued address lines and 8 data registers of binary values [7, p. 170]. The fitness of the 11MX is an error rate for total inputs. The 4EVEN (even-4-parity) problem is to generate a Boolean function which returns T if an even number of its Boolean arguments are T, and otherwise returns NIL [7, p. 158]. The fitness of the 4EVEN is also an error rate for inputs, i.e., the same as that of the 11MX. The ANT problem is the task of navigating an artificial ant so as to find all 89 foods lying along an irregular trail on 32× 32 world [7, pp. 54]. The ant's goal is to traverse the entire trail (thereby eating all of the foods) within a limited energy. The fitness of ANT is the ratio of which the ant could not eat 89 foods. In other words, if the ant could not eat any foods, the fitness if 1.0 whereas if the ant could eat all foods within a limited energy, it is 0.0. For the sake of comparison, all experiments were conducted until a final generation, even if a solution was found during the evolution. Experimental results are shown on the average over twenty runs. Note that the smaller, the better the fitness is.

Table 1 shows the experimental set up. In case of **NORMAL**, crossover points are selected at random. **DD** means using the "depth-dependent" crossover. **RDD** represents the "revised depth-dependent" crossover. These three depth-dependent crossover operators are destructive one. **ND-NORMAL** denotes the "non-destructive normal" crossover. **ND-DD** means using the "non-destructive depth-dependent" crossover. **ND-RDD** represents the non-destructive revised depth-dependent crossover. These three non-destructive crossover operators are mentioned in section 3. Mutation is randomly applied for all settings.

Fig. 4 plots the tree depth, the number of nodes and the number of leaf nodes with generations for the 11MX problem. The used parameters are shown in table 2. According to Fig. 4(a), the tree depth of the non-destructive crossover (**ND-NORMAL, ND-DD** and **ND-RDD**) is smaller than by the destructive

Table 1. Table 1: Experimental Set Up

Setting	Crossover	Mutation
NORMAL	Random	Random
DD	Depth-Dependent	Random
RDD	Revised Depth-Dependent	Random
ND-NORMAL	Random (non-destructive)	Random
ND-DD	Depth-Dependent (non-destructive)	Random
ND-RDD	Revised Depth-Dependent (non-destructive)	Random

Table 2. Parameters for 11MX problem

parameter	value
population size	2000
max of generation	20
max depth for new trees	10
max depth after crossover	15
max mutant depth	3
tree initialization method	grow
selection method	tournament
tournament size	5
crossover function point fraction	0.1
crossover any point fraction	0.6
fitness prop. reproduction fraction	0.2

one (**NORMAL, DD** and **RDD**). The number of (leaf) nodes by the non-destructive crossover is also smaller than by the destructive crossover (see Fig. 4(b) and Fig. 4(c)). Tree growth pattern of the destructive crossover operations (**NORMAL, DD** and **RDD**) is rising until the final generations. However, the non-destructive crossover operations (**ND-NORMAL, ND-DD** and **ND-RDD**) suppress the program growth in the middle of the evolution. These experimental results show that the applicability of non-destructive crossover for combining other crossover techniques. The non-destructive crossover could generate smaller programs on the three GP problems (we show only the tree structures on the 11MX problem because of the page limitation). This suggests that the non-destructive crossover has no problem-dependent characteristics. Our purpose is to suppress the program growth. There are some advantages when the size of generated programs is reduced. For instance, a small program does not require huge computer memory. And the computational time of a small program is smaller than that of a huge program.

Fig. 5(a) and Table 5 give the fitness values and the hits for the 11MX problem. The figure shows that **ND-DD** gave the best performance for the best fitness value. The evolution of **ND-DD** is the fastest among six crossover settings. On the contrary, the evolution of **NORMAL** is the slowest. As for the hits at the final generations, **ND-DD** was also the best among the six cases (see Table 5). We statistically compared non-destructive crossover with destructive crossover using the paired t-test [8]. We could not conclude that **ND-DD** is superior to **DD** in terms of the best fitness value. On the contrary, we have confirmed that **RDD** is superior to **ND-RDD** in terms of the best fitness value (5% of level of significance. see table 6).

Fig. 5(b) plots the results of the 4EVEN problem. The used parameters are shown in table 3. **DD**, **RDD** and **ND-DD** give good performance in terms of the best fitness value. However, **NORMAL**, **ND-NORMAL** and **ND-RDD** fell into local minimum. This is because these three settings generated smaller GP trees than that by the former three settings (i.e., **DD**, **RDD** and **ND-DD**). The solution program was acquired for all twenty runs by means of **DD** (Table 5). According to the paired t statistical test, we have confirmed that **DD** is superior to **ND-DD**, and that **ND-RDD** is superior to **RDD** in terms of the best fitness values (Table 6).

Fig. 5(c) plots the results of the ANT problem. The used parameters are shown in table 4. The performance of **NORMAL** was the best in terms of the best fitness values. On the contrary, **ND-RDD** was the worst in terms of the best fitness value. As can be seen from this figure, all six crossover settings suffered from the into local minimum. The destructive crossover and the non-destructive crossover is same performance on the ANT problem (see Table 5). We could also reduce the GP program size on the ANT problem (we do not show the tree structures on the ANT problem because of the page limitation). In other words, the non-destructive crossover could reduce the GP program size without degrading GP fitness performance. Next section discusses why the depth-dependent crossover (the destructive crossover and the non-destructive crossover) fails in the ANT problem.

Table 3. Parameters for 4EVEN problem

parameter	value
population size	2000
max of generation	24
max depth for new trees	6
max depth after crossover	12
max mutant depth	4
tree initialization method	grow
selection method	tournament
tournament size	5
crossover function point fraction	0.1
crossover any point fraction	0.6
fitness prop. reproduction fraction	0.2

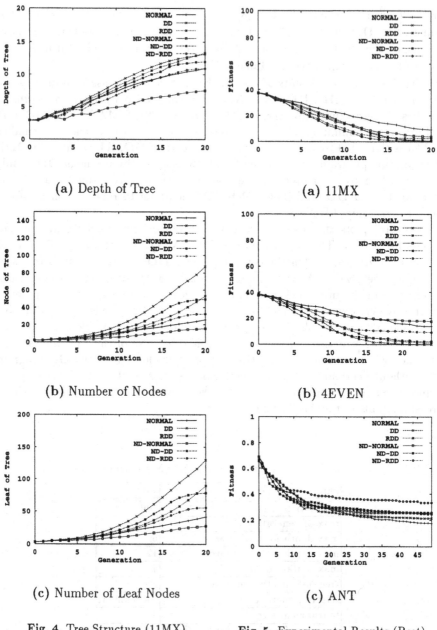

(a) Depth of Tree

(a) 11MX

(b) Number of Nodes

(b) 4EVEN

(c) Number of Leaf Nodes

(c) ANT

Fig. 4. Tree Structure (11MX)

Fig. 5. Experimental Results (Best)

Table 4. Parameters for ANT problem

parameter	value
population size	2000
max of generation	49
max depth for new trees	10
max depth after crossover	17
max mutant depth	3
tree initialization method	grow
selection method	tournament
tournament size	5
crossover function point fraction	0.1
crossover any point fraction	0.6
fitness prop. reproduction fraction	0.2

Table 5. Averaged Numbers of Hits and Standard Deviation Values at the Final Generations over Twenty Runs. Rank is Written in Bold Type Font.

Setting	11MX	4EVEN	ANT
NORMAL	0.00 (0.00) **5**	0.00 (0.00) **6**	0.15 (0.36) **1**
DD	0.70 (0.46) **2**	1.00 (0.00) **1**	0.10 (0.30) **2**
RDD	0.85 (0.36) **1**	0.80 (0.40) **2**	0.00 (0.00) **4**
ND-NORMAL	0.40 (0.49) **4**	0.05 (0.22) **5**	0.05 (0.22) **3**
ND-DD	0.85 (0.36) **1**	0.75 (0.43) **3**	0.05 (0.22) **3**
ND-RDD	0.65 (0.48) **3**	0.15 (0.36) **4**	0.05 (0.22) **3**

Table 6. Statistic t (Best Fitness Value)

Setting	11MX	4EVEN	ANT
ND-NORMAL (against **NORMAL**)	5.14	-1.57	-1.61
ND-DD (against **DD**)	0.28	-2.10	-0.64
ND-RDD (against **RDD**)	-2.82	-5.00	-2.06

5 Discussion

The previous experimental results showed that the **ND-DD** was not always superior to the **DD** in terms of the best fitness value on the 4EVEN problem. However, the **ND-DD** was superior to the **DD** in terms of the average fitness value (see Fig. 6). According to t-test, we have confirmed that **ND-DD** was superior to **DD** in terms of the average fitness value (Fig. 8). This is because most of the individuals "encapsulate junk sub-programs" as a result of the **DD**. "Encapsulation" means the protection of programs from the destructive crossover. "Junk" means a sub-program which does not contribute to generate a solution. A small number of individuals get some benefit of improving its fitness from the destructive depth-dependent crossover. In other words, if the encapsulation works well, a fitness of a program is improved. Thus, the non-destructive crossover is

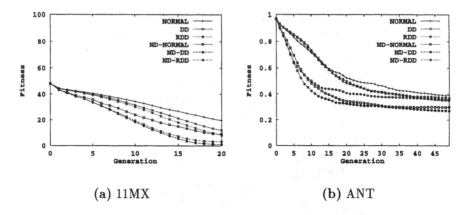

(a) 11MX   (b) ANT

Fig. 6. Experimental Results (Average)

not a greedy method. A greedy method is a search method in which most of the individuals are sacrificed to generate a few elites, which is seen by the depth-dependent crossover. This claim is clarified by the autocorrelation analysis described below. The autocorrelation evaluates the mutual role of the fitness function and the genetic operator in the search process on the landscape [9]. The experiment of the autocorrelation is to generate random programs and then to apply the crossover operation to these programs. We repeated twenty runs of generating random 10,000 individuals to measure the autocorrelation. Used parameters are the same as the former experiments. The autocorrelation ranges from -1 to 1. The closer to 1, the more highly correlated the fitness and the operator are, which means a desirable situation for the adaptive search. According to this analysis, the non-destructive crossover shows a higher correlation than the destructive crossover for the three problems (see Table 7). This result supports our hypothesis, i.e., most of the individuals "encapsulate junk sub-programs" by means of the **DD**.

On the ANT problem, **ND-DD** was the best in terms of the average fitness values. According to the paired t statistical test, we have confirmed that **RDD** was superior to **ND-RDD** in terms of the average fitness values. However, we could not conclude that **ND-DD** was superior to **DD** in terms of the best and the average fitness values (Table 8). There was no critical difference of the fitness between the destructive crossover and the non-destructive crossover.

Table 7. Autocorrelation

Problem	NORMAL	DD	RDD	ND-NORMAL	ND-DD	ND-RDD
11MX	0.38	0.40	0.40	0.83	0.75	0.80
4EVEN	0.47	0.38	0.34	0.93	0.89	0.89
ANT	0.48	0.41	0.41	0.91	0.84	0.83

Table 8. Statistic t (Average)

Setting	11MX	4EVEN	ANT
ND-NORMAL (against **NORMAL**)	12.36	2.22	1.93
ND-DD (against **DD**)	5.15	5.06	1.72
ND-RDD (against **RDD**)	2.20	-0.10	0.41

Crossover promotes the program growth in size [11]. The substantial number of crossover by the non-destructive crossover is smaller than by the destructive crossover. This is because in case of the non-destructive crossover, the parent is copied to the next generation if its fitness is not worse than that of the offspring. Thus, the non-destructive crossover is expected to generate smaller programs than the destructive crossover. In the case of the Boolean concept formation, the relationship between the fitness and the program size is large, and, in general, the large the size, the better the fitness. One reason why the depth-dependent crossover was successful in the Boolean concept formation seems to be the generation of large programs. In case of the non-destructive crossover, in which only a better offspring is kept for the next generation, the program growth in size is prevented so that GP may be easily trapped in a local minimum.

In case of the ANT problem, the performance difference could be explained not by the program size problem, but by the failure of the building block hypothesis. This means that the poor performance was attributed to the characteristics of the depth-dependent crossover. It has been pointed out that "the best class of problems to look for potential building blocks would be the boolean functions. Complex boolean functions often contain sub-functions that can be composed into the larger solutions" [2, p. 16] [5]. Our previous results also confirmed this claim, i.e., the depth-dependent crossover and the non-destructive depth-dependent crossover ware suitable for the problems that can be decomposed into sub-problems.

6 Conclusion

We introduced the non-destructive operation for the depth-dependent crossover and verified the effectiveness in terms of generated program size and fitness values. As a result of experiments, the following points have been made clear:

1. For Boolean problems, the non-destructive depth-dependent crossover created smaller programs than the depth-dependent crossover.
2. For the ANT problem, the performance for the non-destructive crossover was not always better.
3. The non-destructive crossover gave higher correlation between the operator and the fitness than the destructive one.

Previous experimental results have shown that the non-destructive crossover (i.e., **ND-DD** and **ND-RDD**) did not necessarily give better performance than the simple depth-dependent crossover (i.e., **DD** and **RDD**). However, our purpose was to reduce the size of generated programs for GP. The non-destructive crossover generated smaller programs with a slight performance degradation. Considering the benefit of the small size of GP programs, such as the robustness, the reduced computational time, and the small memory requirement, we believe that it pays to apply a non-destructive crossover, especially to a large problem. We are currently working on the theoretical study as to how the depth-dependent crossover accumulates building blocks. The future research will include applying our approach to a real-world problem, such as a robot task [4], and coping with a dynamic environment. We are also studying the mechanism of the self-tuning of the depth selection ratio.

References

1. Aho, A., Hopcroft, J. and Ullman, J. *The Design and Analysis of Computer Algorithms*, Addison-Wesley, 1974
2. Angeline, J. *Subtree Crossover: Building Block Engine or Macromutation?*, In Koza, J., Deb, K., Dorigo, M., Fogel, D. Garzon, M., Iba, H. and Riolo, R. editors, *Proceedings of the Second Annual Conference Genetic Programming 1997 (GP97)*, pages 9–17, MIT Press, 1997
3. Iba, H., deGaris, H. and Sato, T. *Genetic Programming using a Minimum Description Length Principle*, In Kinnear, Jr. K. editor, *Advances in Genetic Programming* , pages 265–284, MIT Press, 1994
4. Ito, T. and Iba, H. and Kimura, M. *Robustness of Robot Programs Generated by Genetic Programming*, In Koza, J., Goldberg, D., Fogel, D. and Riolo, R. editors, *Proceedings of the First Annual Conference Genetic Programming 1996 (GP96)*, pages 321–326, MIT Press, 1996
5. Ito, T. and Iba, H. and Sato, S. *Depth-Dependent Crossover for Genetic Programming*, In Proceedings of the 1998 IEEE International Conference on Evolutionary Computation (ICEC'98), 1998
6. Kinnear, Jr. K. *Generality and Difficulty in Genetic Programming: Evolving a Sort*, In Proceedings of 5th International Joint Conference on Genetic Algorithms MIT press, 1993
7. Koza, J. *Genetic Programming: On the Programming of Computers by Natural Selection*, MIT press, 1992
8. Rudolf, F. and Wilson, W. *Statistical Methods*, Academic Press, Inc. 1992
9. Slavov, V and Nikolaev, N *Fitness Landscapes and Inductive Genetic Programming*, In *In Smith, G. editor, Third International Conference on Artificial Neural Networks and Genetic Algorithms (ICANNGA'97)*, Springer-Verlag, Vienna, 1997
10. Soule, T., Foster, J. and Dickinson, J. *Code Growth in Genetic Programming*, In Koza, J., Goldberg, D., Fogel, D. and Riolo, R. editors, *Proceedings of the First Annual Conference Genetic Programming 1996 (GP96)*, pages 215–223, MIT Press, 1996
11. Soule, T. and Foster, J. *Code Size and Depth Flows in Genetic Programming*, In Koza, J., Deb, K., Dorigo, M., Fogel, D. Garzon, M., Iba, H. and Riolo, R. editors, *Proceedings of the Second Annual Conference Genetic Programming 1997 (GP97)*, pages 313–320, MIT Press, 1997

Grammatical Evolution : Evolving Programs for an Arbitrary Language

Conor Ryan, JJ Collins & Michael O Neill

Dept. Of Computer Science And Information Systems
University of Limerick
Ireland
{Conor.Ryan|J.J.Collins|Michael.ONeill}@ul.ie

Abstract. We describe a Genetic Algorithm that can evolve complete programs. Using a variable length linear genome to govern how a Backus Naur Form grammar definition is mapped to a program, expressions and programs of arbitrary complexity may be evolved. Other automatic programming methods are described, before our system, *Grammatical Evolution*, is applied to a symbolic regression problem.

1 Introduction

Evolutionary Algorithms have been used with much success for the automatic generation of programs. In particular, Koza's [Koza 92] Genetic Programming has enjoyed considerable popularity and widespread use. Koza's method originally employed Lisp as its target language, and others still generate Lisp code. However, most experimenters generate a homegrown language, peculiar to their particular problem.

Many other approaches to automatic program generation using Evolutionary Algorithms have also used Lisp as their target language. Lisp enjoys much popularity for a number of reasons, not least of which is the property of Lisp of not having a distinction between programs and data. Hence, the structures being evolved can directly be evaluated. Furthermore, with reasonable care, it is possible to design a system such that Lisp programs may be safely crossed over with each other and still remain syntactically correct.

Evolutionary Algorithms have been used to generate other languages, by using a grammar to describe the target language. Researchers such as Whigham [Whigham 95] and Wong and Leung's LOGENPRO system [Wong 95] used context free languages in conjuntion with GP to evolve code. Both systems exploited GP's use of trees to manipulate parse trees, but LOGENPRO did not explicitly maintain parse trees in the population, and so suffered from some ambiguity when trying to generate a tree from a program. Whigham's work did not suffer from this, and had the added advantage of allowing an implementor to bias [Whigham 96] the search towards parts of the grammar.

Another attempt was that of Horner [Horner 96] who introduced a system called Genetic Programming Kernel (GPK), which, similar to standard GP, employs trees to code genes. Each tree is a derivation tree made up from the BNF definition. However, GPK has been criticised [Paterson 97] for the difficulty associated with generating the first generation - considerable effort must be put into ensuring that all the trees represent valid sequences, and that none grow without bounds. GPK has not received widespread usage.

An approach which generates C programs directly, was described by Paterson [Paterson 97]. This method was applied to the area of evolving caching algorithms in C with some success, but with a number of problems, most notably the problem of the chromosomes containing vast amounts of introns. This method will be more fully discussed in section 3. We describe a different approach to using BNF definitions, and develop a system that evolves individuals containing no introns. This system can be used to evolve programs in any language. We adopt the approach that the genotype must be mapped to the phenotype [Keller 96] [Ryan 97a] [Ryan 97b] [Gruau 94] rather than treating the actual executable code as data. In this respect we diverge from Koza's approach, but with the result that the individuals tend to be much smaller.

2 Backus Naur Form

Backus Naur Form (BNF) is a notation for expressing the grammar of a language in the form of production rules. BNF grammars consist of terminals, which are items that can appear in the language, i.e $+, -$ etc. and non-terminals, which can be expanded into one or more terminals and non-terminals. A grammar can be represented by the tuple, $\{N, T, P, S\}$, where N is the set of non-terminals, T the set of terminals, P a set of production rules that maps the elements of N to T, and S is a start symbol which is a member of N. For example, below is a possible BNF for a simple expression, where

$$N = \{expr, op, pre\_op\}$$

$$T = \{Sin, Cos, Tan, Log, +, -, /, *, X, ()\}$$

$$S = < expr >$$

And P can be represented as:

(1) <expr> ::= <expr> <op> <expr>   (A)
             | (<expr> <op> <expr>)   (B)
             | <pre-op> (<expr>)   (C)
             | <var>   (D)

(2) <op> ::= + (A)
           | - (B)
           | / (C)
           | * (D)

(3) <pre-op> ::= Sin (A)
               | Cos (B)
               | Tan (C)
               | Log (D)

(4) <var> ::= X

Unlike a Koza-style approach, there is no distinction made at this stage between what he describes as functions (operators in this sense) and terminals (variables in this example), however, this distinction is more of an implementation detail than a design issue.

Whigham [Whigham 96] also noted the possible confusion with this terminology and used the terms **GPFunctions** and **GPTerminals** for clarity. We will also adopt this approach, and use the term **terminals** with its usual meaning in grammars.

Table 1 summarizes the production rules and the number of choices associated with each. When generating a sentence for a particular language, one must choose carefully which productions are to be used, as, depending on the choices made, a sentence may be quite different from the desired one, possibly even of a different length.

Rule no.	Choices
1	4
2	4
3	4
4	1

Table 1. The number of choices available from each production rule.

We propose to use a genetic algorithm to control what choices are made at each juncture, thus allowing the GA to control what production rules are used. In this manner, a GA can be used to generate any manner of code in any language.

3   Genetic Algorithm for Developing Software

Genetic Algorithm for Developing Software (GADS) as described by [Paterson 97] uses fixed length chromosomes which encode which production rules are to be applied. If, when interpreting a gene, it doesn't make syntactic sense, e.g. trying to apply rule 2.A as the first production, or when no operator is available, it is ignored.

If, at the end of the chromosome, there are gaps in the expression, i.e. non-terminals which did not have any terminal chosen for them, a default value is inserted. The default value must be tailored for each production rule. In [Paterson 97] it was suggested that an empty production is the suitable approach, however, this is not always possible. Consider the BNF above, rules 2 and 3 must produce an operator of some description, otherwise the entire expression would be compromised. Thus, an arbitrary decision must be taken about which rule should be used as the default. If, for example, an individual was generated that had the non terminal $< op >$ in it, with all of the genes exhausted, one must decide which of the rules $2A..2D$ should be used as a default rule. Clearly, an unfortunate or misguided choice could harm the evolution of a population.

4   Grammatical Evolution

The GADs approach suffers from a number of drawbacks. In particular, as the number of productions grows, the chance of any particular production being chosen by a gene reduces. Paterson's suggestion to combat this was to simply increase the population size or the genome size. Another suggestion was to duplicate certain production rules in the BNF.

This results in a serious proliferation of introns, consider (using our own notation, not Patersons) the following chromosome using the productions from above.

| 1D | 1A | 2B | 3B | 3A | .... | 4A |

Because of the initial rule that was used, there is only one possible choice rule to apply, and any other rule that occurs in the genome must be ignored. This can lead to a situation where the majority of genome consists of introns.

Instead of coding the transitions, our system codes a set of pseudo random numbers, which are used to decide which choice to take when a non terminal has one or more outcomes. A chromosome consists of a variable number of binary genes, each of which encodes an 8 bit number.

| 220 | 203 | 17 | 3 | 109 | 215 | 104 | 30 |

Table 2. The chromosome of an individual. Each gene represents a random number which can be used in the translation from genotype to phenotype.

Consider rule #1 from the previous example:

```
(1) <expr> ::= <expr> <op> <expr> | ( <expr> <op> <expr> ) |
               <pre-op> ( <expr> ) | <var>
```

In this case, the non-terminal can produce one of four different results. our system takes the next available random number from the chromosome to decide which production to take. Each time a decision has to be made, another pseudo random number is read from the chromosome. and in this way, the system traverses the chromosome.

In natural biology, there is no direct mapping from gene to physical expression [Elseth 95]. When genes are expressed they generate proteins, which, either independantly or, more commonly in conjunction with other proteins, created by other genes, affect physical traits. We treat each transition as a protein, on their own, each transition cannot generate a physical trait. However, when other proteins are present, physical traits can be generated. Moreover, while a particular gene always generates the same protein, the physical results depend on the other proteins that are present immediately before and after.

Consider the individual in Table 2. The fourth gene generates the number 3, which, in our system, is analgous to the *protein* 3. Notice that this will be generated regardless of where the gene appears on the chromosome. However, it may have slightly different effects depending on what other proteins have been generated previously. The following section describes the mapping from genotype to phenotype in detail.

It is possible for individuals to run out of genes, and in this case there are two alternatives. The first is to declare the individual invalid and punish them with a suitably harsh fitness value; the alternative is to wrap the individual, and reuse the genes. This is quite an unusual approach in EAs, as it is entirely possible for certain genes to be used two or more times. Each time the gene is expressed it will always generate the same protein, but depending on the other proteins present, may have a different effect. The latter is the more biologically plausible approach, and often occurs in nature. What is crucial, however, is that each time a particular individual is mapped from its genotype to its phenotype, the same output is generated. This is because the same choices are made each time.

To complete the BNF definition for a C function, we need to include the following rules with the earlier definition:

```
<func> ::= <header>

<header> ::= float symb(float X) { <body> }

<body> ::= <declarations><code><return>

<declarations ::= float a;

<code> ::= a = <expr>;

<return> ::= return (a);
```

Notice that this function is limited to a single line of code. However, this is because of the nature of the problem, the system can easily generate functions which use several lines of code. Specifically, if the rule for code was modified to read:

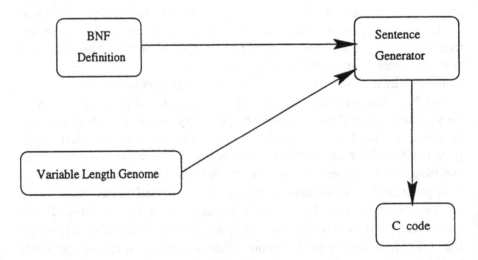

Fig. 1. The Grammatical Evolution System

```
<code> ::= <line>;  | <line>; <code>
```

```
<line> ::= <var> = <expr>
```

then the system could generate functions of arbitrary length.

4.1 Example Individual

Consider an individual made up of the following genes (expressed in decimal for clarity) :

220	203	17	3	109	215	104	30

These numbers will be used to look up the table in Section 2 which describes the BNF grammar for this particular problem. The first few rules don't involve any choice, so all individuals are of the form:

```
float symb(float x)

  {
  a = <expr>;
  return(a);
  }
```

Concentrating on the <expr> part, we can see that there are four productions to choose from. To make this choice, we read the first gene from the chromosome, and use it to generate a protein in the form of a number. This number will then be used to decide which production rule to use, thus s we have 220 MOD 4 = 0 which means we must take the first production, namely, 1A. We now have the following

```
<expr> <op> <expr>
```

Notice that if this individual is subsequently wrapped, the first gene will still produce the protein 220. However, depending on previous proteins, we may well be examining the choice of another rule, possibly with a different amount of choices. In this way, although we have the same protein it results in a different physical trait.

Continuing with the first expression, a similar choice must be made, this time using 203 *MOD* 4 = 3, so the third choice is used, that is 1C.

```
<var> <op> <expr>
```

There is no choice to be made for var, as there is only one possible outcome, so *X*, is inserted. Notice that no number is read from the genome this time.

```
X <op> <expr>
```

The mapping continues, as summarized in table 3, until eventually, we are left with the following expression:

```
X + Sin ( X )
```

Notice how not all of the genes were required, in this case the extra genes are simply ignored.

5   Operators

Due to the fact that genes are mapped from genotype to phenotype there is no need for problem specific genetic operators, and GE employs all the standard operators of Genetic Algorithms. There are however, two new operators **Prune** and **Duplicate** , which are peculiar to GE.

Expression	Rule	Gene	Number Generated
<expr>	n/a		
<expr><op><expr>	1A	220	0
<var><op><expr>	1D	203	3
X<op><expr>	4	n/a	
X+<expr>	2A	17	1
X+pre_op(<expr>)	2C	19	3
X+Sin(<expr>)	3A	109	1
X+Sin(<var>)	1D	215	3
X+Sin(X)	4	n/a	

Table 3. Mapping a genome onto an expression

5.1 Duplicate

Gene duplication involves making a copy of a gene or genes [Elseth 95], and has been used many times in Evolutionary computation [Schutz 97]. Duplicated genes can lie adjacent to the original gene, or can lie at a completely different region. Repeated gene duplications are presumed to have given rise to the human hemoglobin gene family. The hemoglobin protein is involved with transporting oxygen around the body through the bloodstream. The original hemoglobin gene duplicated to form an α - globin, and β - globin genes. Each of these underwent further duplications resulting at present in four variations on the α - globin gene, and five of the β - globin gene. Different combinations of these genes are expressed during development of the human, resulting in the hemoglobin having different binding properties for oxygen. The presence of a copy of a gene can therefore have benefits. Biologically the results of gene duplications can be to supply genetic material capable of:

(i) Acquiring lethal mutation(s). If two copies of a gene are present in a genome, and one happens to be mutated such that it's protein product is no longer functional having the second copy means that this probability of harmful mutation is reduced.

(ii) Evolving new functions. If a mutation is favourable it may result in an altered protein product, perhaps with new functionality, without changing the function of the original gene's protein product.

(iii) Duplicating a protein in the cell, thus producing more of it. The increased presence of a protein in a cell may have some biological effects. Goldberg [Goldberg 89] stated that using operators such as duplication in a variable-length genotype enables them to solve problems by combining relatively short, well-tested building blocks to form longer, more complex strings that increasingly cover all features of a problem.

The duplication operator used in GE is a multiple gene duplication operator. The number of genes to be duplicated is picked randomly. The duplicated genes are placed into the position of the last gene on the chromosome, if the (X * X) part of the expression were duplicated, this could result in a useful extension to the genome, especially if a multiplication operator happend to be between the original expressions genes and the duplicated one. Gene duplication in GE is essentially analogous to producing more copies of a gene(s) to increase the presence of a protein(s) in the cell.

5.2 Pruning

Individuals in GE don't necessarily use all their genes. For example, to generate the expression $X * X + X$ only five genes are required, but an individual that generates this expression could well be made up of many more. The remaining genes are introns and serve to protect the crucial genes from being broken up in crossover. While this certainly serves the selfish nature of genes in that it increases the longevity of a particular combination, it is of no advantage to us. In fact, it would make sense for an individual to be as long as possible, to prevent its combination from being disrupted. The fact that such disruption could be crucial to the evolution of the population as a whole is of consequence to the individual. Figure 2 illustrates the effect of too many introns.

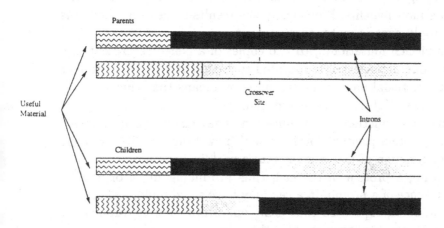

Fig. 2. Crossover being hampered by introns

To reduce the number of introns, and thus increase the likelihood of beneficial crossovers, we introduce the **prune** operator. **Prune** is applied with a probability to any individuals that don't express all of their genes.. Any genes not used in the genotype to phenotype mapping process are discarded.

The effects of pruning are dramatic; FASTER; BETTER CROSSOVER.

6   The Problem Space

As proof of concept, we have applied our system to a symbolic regression problem used by Koza [Koza 92]. The system is given a set of input and output pairs, and must determine the function that maps one onto the other. The particular function Koza examined is $X^4 + X^3 + X^2 + X$ with the input values in the range $[-1..+1]$. Table 4 contains a tableau which summarizes his experiments.

Objective :	Find a function of one independant variable and one dependant variable, in symbolic form that fits a given sample of 20 (x_i, y_i) data points, where the target functions is the quartic polynomial $X^4 + X^3 + X^2 + X$
GPTerminal Set:	X (the independant variable)
GPFunction Set	$+, -, *, \%, \sin, \cos, \exp, \log$
Fitness cases	The given sample of 20 data points in the interval $[-1, +1]$
Raw Fitness	The sum, taken over the 20 fitness cases, of the error
Standardised Fitness	Same as raw fitness
Hits	The number of fitness cases for which the error is less than 0.01
Wrapper	None
Parameters	$M = 500$, $G = 51$

Table 4. A Koza-style tableau

The production rules for $< expr >$ are given above. As this and subsequent rules are the only ones that require a choice, they are the ones that will be evolved.

We adopt a similar style to Koza of summarizing information, using a modified version of his *tableau*. Notice how our *terminal operands* and *terminal operators* are analgous to *GPTerminals* and *GPfunctions* respectively.

6.1 Results

GE consistently discovered the target function when both the duplication and pruning operators were employed. Examination of potential solutions from the other runs showed they tended to fixate on operators such as Sin or Exp as these gene rated a curve similar to that of the target function.

Space constraints prevent us from describing the nature of individuals generated by GE as it was constructing the correct solution. However, the system exploits the variable length genome by concentrating on the start of the string initially. Although the target function can be represented by eighteen distinct genes, there are several other possibilities which, while not perfectly fit, are considerably shorter. Individuals such as $X^2 + X$ generate curves quite comparable to the target, using a mere five genes.

Objective :	Find a function of one independant variable and one dependant variable, in symbolic form that fits a given sample of 20 (x_i, y_i) data points, where the target functions is the quartic polynomial $X^4 + X^3 + X^2 + X$
Terminal Operands:	X (the independant variable)
Terminal Operators	The binary operators $+, *, /$, and $-$ The unary operators Sin, Cos, Exp and Log
Fitness cases	The given sample of 20 data points in the interval $[-1, +1]$
Raw Fitness	The sum, taken over the 20 fitness cases, of the error
Standardised Fitness	Same as raw fitness
Hits	The number of fitness cases for which the error is less than 0.01
Wrapper	Standard productions to generate C functions
Parameters	$M = 500, G = 51$

Table 5. Grammatical Evolution Tableau

Typically, individuals of this form were discovered early in a run, and contained valuable gene sequences, particularly of the form $X * X$ which, if replicated could subsequently be used to generate individuals with higher powers of X. GE is subject to problems of dependencies similar to GP [O'Reilly 97], i.e. the further from the root of a genome a gene is, the more likely its function is to be affected by other genes.

By biasing individuals to a shorter length, they were encouraged to discover shorter, albeit less fit expressions early in a run, and then generate progressively longer genomes and hence increasingly complex functions in subsequent generations.

7 Conclusions

We have described a system, Grammatical Evolution (GE) that can map a binary genotype onto a phenotype which is a high level program. Because our mapping technique employs a BNF definition, the system is language independant, and, theoretically can generate arbitrarily complex functions.

We believe GE to have closer biological analogies to nature than GP, particularly with its use of linear genomes and the manner in which it uses proteins to affect traits. Other schemes tend to use either the direct evolution of physical traits or the use of simple one to one mappings from genotype to phenotypic traits.

This paper serves an introduction to GE. We believe there is great promise to the system, and that the ability to evolve any high level language would surely be a huge boost for Evolutionary Algorithms, both in terms of increased usage and acceptance, and in terms of the complexity of the problems it can tackle.

Using BNF definitions, it is possible to evolve multi line functions; these functions can even declare local variables and use other high level constructs such as loops. The next step is to apply GE to a problem that requires such constructs.

Although GE is quite different from the functional nature GP, in both its structure and its aims, there is still quite a lot of similarity. We hope that GE will be able to benefit from the huge amount of work already carried out on GP, particularly in the fundamentals, i.e. ADFS, indexed memory, etc. many of which can be coded in BNF with relative ease.

References

[Elseth 95] Elseth Gerald D., Baumgardner Kandy D. Principles of Modern Genetics.
West Publishing Company

[Goldberg 89] Goldberg D E, Korb B, Deb K. Messy genetic algorithms: motivation, analysis, and first results.
Complex Syst. 3

[Gruau 94] Gruau, F. 1994. *Neural Network synthesis using cellular encoding and the genetic algorithm.* PhD Thesis from Centre d'etude nucleaire de Grenoble, France.

[Horner 96] Horner, H *A C++ class library for GP.* Vienna University of Economics.

[Keller 96] Keller, R. & Banzhaf, W. 1996. GP using mutation, reproduction and genotype-phenotype mapping from linear binary genomes into linear LALR phenotypes. In *Genetic Programming 1996*, pages 116-122. MIT Press.

[Koza 92] Koza, J. 1992. *Genetic Programming*. MIT Press.

[O'Reilly 97] O'Reilly, U. 1997. The impact of external dependency in Genetic Programming Primitives. In *Emerging Technologies 1997*, pages 45-58. University College London. *To Appear.*

[Paterson 97] Paterson, N & Livesey, M. 1997. Evolving caching algorithms in C by GP. In *Genetic Programming 1997*, pages 262-267. MIT Press.

[Ryan 97a] Ryan, C. & Walsh P. 1997. The Evolution of Provable Parallel Programs. In *Genetic Programming 1996*, pages 406-409. MIT Press.

[Ryan 97b] Ryan, C. 1997. Shades - A Polygenic Inheritance scheme. In *Proceedings of Mendel '97*, pages 140-147. PC-DIR, Brno, Czech Republic.

[Schutz 97] Schutz, M. 1997. Gene Duplication and Deletion, in the Handbook of Evolutionary Computation. (1997) Section C3.4.3

[Whigham 95] Whigham, P. 1995. Inductive bias and genetic programming. In *First International Conference on Genetic Algorithms in Engineering Systems: Innovations and Applications*, pages 461-466. UK:IEE.

[Whigham 96] Whigham, P. 1996. Search Bias, Language Bias and Genetic Programming. In *Genetic Programming 1996*, pages 230-237. MIT Press.

[Wong 95] Wong, M. and Leung, K. 1995. Applying logic grammars to induce subfunctions in genetic prorgramming. In *Proceedings of the 1995 IEEE conference on Evolutionary Computation*, pages 737-740. USA:IEEE Press.

Genetic Programming Bloat
with Dynamic Fitness

W. B. Langdon and R. Poli

School of Computer Science, University of Birmingham, Birmingham B15 2TT, UK
{W.B.Langdon,R.Poli}@cs.bham.ac.uk http://www.cs.bham.ac.uk/~wbl, ~rmp
Tel: +44 (0) 121 414 4791, Fax: +44 (0) 121 414 4281

Abstract. In artificial evolution individuals which perform as their parents are usually rewarded identically to their parents. We note that Nature is more dynamic and there may be a penalty to pay for doing the same thing as your parents. We report two sets of experiments where static fitness functions are firstly augmented by a penalty for unchanged offspring and secondly the static fitness case is replaced by randomly generated dynamic test cases. We conclude genetic programming, when evolving artificial ant control programs, is surprisingly little effected by large penalties and program growth is observed in all our experiments.

1 Introduction

The tendency for programs in genetic programming (GP) populations to grow in length has been widely reported [Tac93; Tac94; Ang94; Tac95; Lan95; NB95; SFD96]. In our previous work on this phenomenon (referred to as "bloat") [LP97; Lan97; Lan98b; LP98a] we have investigated the effect of commonly used fitness functions with a variety of genetic algorithm, population based and non-population based search techniques. These experiments have shown that bloat is not a unique phenomenon to genetic programming and we argue that it is inherent in discrete variable length representations using a simple scalar static fitness function. [NB95; Lan98b] suggest that non-performance effecting code (sometimes referred to as "introns") can contribute to bloat in population based search techniques, such as GP. Our explanations stress the role of simple static scalar fitness functions in selecting children which behave in the same way as their parents. In this paper we broaden our investigation to consider fitness functions which avoid selecting such children and dynamic fitness functions (where the test case changes every generation).

We continue to use the well known genetic programming bench mark problem of evolving control programs to guide an artificial ant along an intermittent trail of food pellets. The first experiments use the well known Santa Fe trail [Koz92], while this is replaced in the second set of experiments by randomly generated trails which are changed each generation (i.e. the population "sees" each trail only once). In these experiments the effects of changing the fitness function are studied.

In Sect. 2 we briefly describe the artificial ant problem and the genetic programming system used to solve it. In Sect. 3 we describe the fitness functions used in the two sets of experiments and introduce the penalty for copying the behaviour of ancestors. Our results are given in Sects. 4 and 5, which are followed by our conclusions in Sect. 6.

2 The Artificial Ant Problem

The artificial ant problem is described in [Koz92, pages 147–155]. It is a well studied problem and was chosen as it has a simple fitness function. [LP98b] shows it is a difficult problem for GP, simulated annealing and hill climbing but has many of the properties often ascribed to real world problems. Briefly the problem is to devise a program which can successfully navigate an artificial ant along a twisting trail on a square 32 × 32 toroidal grid. The program can use three operations, Move, Right and Left, to move the ant forward one square, turn to the right or turn to the left. Each of these operations takes one time unit. The sensing function IfFoodAhead looks into the square the ant is currently facing and then executes one of its two arguments depending upon whether that square contains food or is empty. Two other functions, Prog2 and Prog3, are provided. These take two and three arguments respectively which are executed in sequence.

The evolutionary system we use is identical to [LP97] except the limit on the size of programs has been effectively removed by setting it to a very large value. The details are given in Table 1, parameters not shown are as [Koz94, page 655]. Note in these experiments we allow the evolved programs to be far bigger than required to solve the problem. (The smallest solutions comprise only 11 nodes [LP98b]).

Table 1. Ant Problem

Objective:	Find an ant that follows food trails
Terminal set:	Left, Right, Move
Functions set:	IfFoodAhead, Prog2, Prog3
Fitness cases:	The Santa Fe trail or randomly generated trails
Fitness:	Food eaten less "plagiarism" penalty
Selection:	Tournament group size of 7, non-elitist, generational
Wrapper:	Program repeatedly executed for 600 time steps.
Population Size:	500
Max program size:	no effective limit
Initial population:	Created using "ramped half-and-half" with a max depth of 6
Parameters:	90% crossover, 10% reproduction, no mutation
Termination:	Maximum number of generations G = 50

3   Fitness Functions

3.1   Santa Fe Trail

The artificial ant must follow the "Santa Fe trail", which consists of 144 squares with 21 turns. There are 89 food units distributed non-uniformly along it. Each time the ant enters a square containing food the ant eats it. The amount of food eaten within 600 time units is the score of the control program.

3.2   Random Trails

In the second set of experiments the fixed Santa Fe trail is replaced by 50 randomly generated trails each containing 80 food items. Each generation is tested on a different trail, however the order of the trails is the same in each run. (The test case is available via anonymous ftp node `ftp.cs.bham.ac.uk` directory `pub/authors/W.B.Langdon/gp-code` in file `dynamic.trl` revision 1.11). Each trail is created by appending (in randomly chosen orientations) 20 randomly chosen trail fragments each containing 4 food pellets. We use the 17 fragments shown in Fig. 1.

A uniform choice from these 17 fragments appeared to produce trails which were too difficult. Therefore, like the Santa Fe trail, the randomly produced trails were made easier at the start. This was implemented by increasing the chance of selecting the lower numbered fragments (as they have smaller gaps in the trail). In detail: 1) start at the origin facing along the x-axis with $n = 0$; 2) select a trail fragment uniformly from fragments numbered $1 \ldots n/2+1$ when n is less than 9 and uniformly from the whole set when it is bigger; 3) the chosen fragment is then rotated and/or reflected into a random orientation from those available (see Fig. 1). Make sure the transformation is compatible with the current direction; 4) the fragment is then appended to the trail, possibly changing the current direction and n is incremented; 5) unless there are already 20 fragments in the trail, go back to (2) and select another fragment. Once the trail is complete, it is checked to see it does not cross the start position and does not fold back over itself (i.e. food pellets are not closer than 2 grid squares, unless they are on the same part of the trail). If either check fails, the trail is discarded and a new one is created.

In practice it is difficult to create a contiguous winding trail of 80 food pellets in a 32×32 grid without it overlapping itself. Therefore a toroidal grid of 300×300 was used. The time available to the ant to transverse each trail is calculated as it is created. The time allowed is five plus the sum of the time allocated to each of the fragments it contains (see Fig. 1) plus a further five, to allow the ant to do some additional local searching, for each occasion when fragments at different orientations are used (i.e. where a new bend is introduced). In practice this makes the problem very difficult. In several runs the best programs evolved showed true trail following abilities but were marginally too inefficient to follow all the trails completely within the time limit.

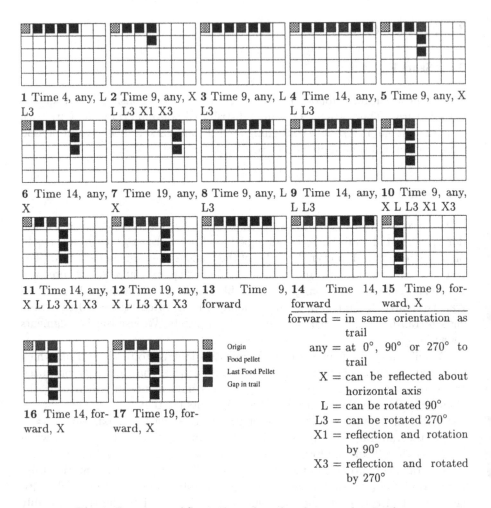

Fig. 1. Fragments of Santa Fe trail used to form random trails.

3.3 Plagiarism Penalty

In both sets of experiments (i.e. on the Santa Fe and on the random trails) runs were conducted with a range of fitness plagiarism penalties. The plagiarism penalty is applied to programs which, when run on the same test as their first parent, have the same score as that parent. (The first parent is defined to be the one from which they inherit their root node). E.g. when using the Santa Fe trail, fitness is reduced if a program causes the ant to eat the same number of food pellets as the program's first parent did. In the case of the dynamic test case, the program must be run both on its own test and the test for the previous generation, i.e. the test its parents where run on.

The smallest penalty (referred to as -0.5) causes the fitness to be reduced by half a food pellet. The other penalties reduce a program's fitness by a fixed fraction. The highest penalty (100%) sets the fitness to zero. As tournament selection is used, the effect the penalty has on the number of children given to each program is complicated, however when the penalty is large compared to the spread of scores within the population, even the best program has little chance of producing children for the next generation.

In many of the results given in the following sections the data falls into a low penalty group (20% or less) and a high penalty group (50% or more). For example in Fig. 3 the average sizes of programs in the population with penalties of 0%, -.5 and 20% lie close to each other, as do 50%, 80%, 95% and 100%, and the two groups are clearly separated. We can estimate the mean plagiarism penalty (when applied) as its size times the mean score in the population. The separation between runs where the penalty is effective and the others corresponds to when this estimate is bigger than the variation in program score across the population (as measured by its standard deviation). That is the plagiarism penalty seems to have little effect if it smaller than the existing variation in programs' scores before it is applied.

4 Santa Fe Trail Results

The results on the Santa Fe Trail are based on ten independent runs with each plagiarism penalty setting. The same ten initial populations are used with each plagiarism value.

From Fig. 2 we can see even the highest plagiarism penalty has only a little depressing effect on the maximum score in the population. Surprisingly the effect on the average program score is more obvious. Suppressing programs which copy their parents appears to considerably reduce the proportion of the best program to the rest of the population, thereby increasing the gap between the mean and maximum raw score. As expected, once the rate of finding higher scoring programs drops (about generation 20, 10,000 programs generated) the size of programs increases and the population bloats, cf. Fig. 3. The plagiarism penalty is unable to prevent this but appears to slow growth, in that when the penalty is 50% or more, by the end of the run programs are on average only half the size of those created in runs with lower penalties.

As expected near the end of runs with low penalty the maximum score in the population remains fixed for many generations. In contrast at the ends of runs with high penalties it varies rapidly. (It changes, either increases or decreases, on average every 2.3 generations in the last ten generations with a penalty of 100% and only increases once in the same period in the ten runs with no penalty. This difference is not obvious in Fig. 2 as it plots data averaged over all ten runs).

With low penalties the population converges in the sense that typically about 90% of programs in the final population have the same score. They are descended, via their first parents, from the same individual. The founding individual has the same score as them, as do its descendants in the genetic lines connecting it to them. In contrast runs with penalties of 20% or more don't appear to converge like this and typically there are only 1 or 2 programs with the highest score in the population.

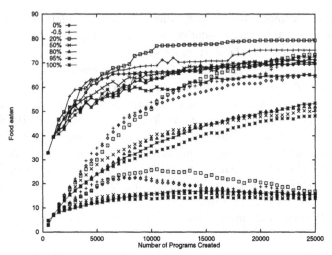

Fig. 2. Evolution of maximum, population mean and standard deviation of food eaten on Santa Fe trail as plagiarism penalty is increased. Means of 10 runs.

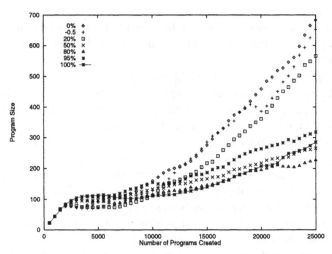

Fig. 3. Evolution of population mean program size on Santa Fe trail as plagiarism penalty is increased. Means of 10 runs.

Further evidence for the lack of convergence is contained in Fig. 4 which indicates where the penalty is low the population evolves so that it is quickly dominated by programs which have the same score as their first parent. While with a high penalty this fraction of the programs increases only slightly and remains near its value in the first few generations. Looking back to earlier generations we see the same picture. With no penalty, typically about 90% of the population has the same score as its first grandparent (i.e. the one it inherited its root from). While with higher penalties it is in the region of 3%. Note this convergence is not the same as convergence in linear genetic algorithms, the population variety is high. Without a plagiarism penalty the fraction of different programs in the population rises to lie near 99% by the end of the run. While with a plagiarism penalty of 50% or more it is still high and reaches about 95% by generation 50. (The slight difference between these figures may be simply due to the larger size of programs when the penalty is low, cf. Fig. 3). Studying the successful crossovers, i.e. those that produced offspring which cause more food to be eaten than their (first) parent, shows runs without a penalty converge in that crossover seldom makes improvement after generation 10 (5,000 programs created) (in the ten run there were only 40 improvements after generation 25, compared to a total of 3,451). Whereas with a 100% penalty the population remains "on the boil" with parents of lower score being selected and crossover continuing to make improvements on them until the end of the run. (E.g. In the first run with a 100% penalty there were 3,098 such crossovers, about 1,730 after generation 25).

Another aspect of the convergence of runs with low penalties is individuals with the same score are selected to be parents. E.g. at the end of all but one run with no penalty, all parents had identical scores. With high penalties there is more variation (on average 29.7 different scores acting as first parents to children in the final population of runs with a 100% penalty).

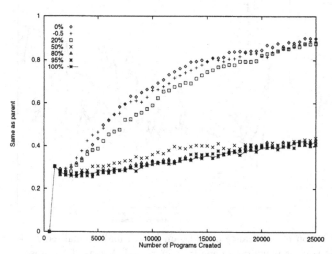

Fig. 4. Evolution of proportion of population with same score as first parent on Santa Fe trail as plagiarism penalty is increased. Means of 10 runs.

4.1 Correlation of Fitness and Program Size

In GP program size is inherited and so we can apply Price's Covariance and Selection Theorem [Pri70; LP98a] to it. Provided our genetic operators are unbiased with respect to length, the expected change in mean program length from one generation to the next is given by the covariance between program length and normalised fitness in the previous generation. (Normal GP crossover is unbiased provided size or depth restrictions don't effect the population. I.e. children produced by crossover are on average the same size as their parents [LP97]). Figure 5 shows in the first few generations there is a strong covariance between length and fitness. We suggest this is because in the initial random populations long programs tend to do better simply because they are more likely to contain useful primitives such as Move. In the next few generations strong selection acts to remove useless programs and consequently the covariance falls. The plagiarism penalty appears to dilute this effect of selection so the covariance in high penalty runs takes more generations to fall. There is then a period of about eight generations during which crossover finds many improved solutions and covariance remains small after which the normal increase is seen in low penalty runs and the populations bloat. Figure 5 shows strong plagiarism penalties appear to prevent the covariance increasing towards the end of the runs so reducing the bloat seen with low penalties (cf. Fig. 3). However the covariance remains positive and some increase in programs size is seen even with the highest penalty.

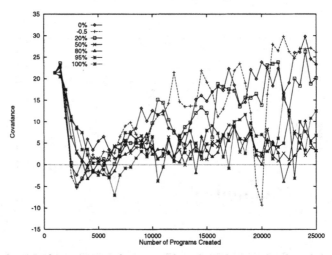

Fig. 5. Evolution of covariance of program length and normalised rank based fitness on Santa Fe trail as plagiarism penalty is increased. Means of 10 runs.

Figure 6 plots the correlation coefficient of program size and amount of food eaten. (Correlation coefficients are equal to the covariance after it has been normalised to lie in the range $-1 \ldots +1$. By considering food eaten we avoid the intermediate stage of converting program score to expected number of children required when applying Price's Theorem and exclude the plagiarism penalty). Figure 6 shows in all cases there is a positive correlation between program score (rather than fitness) and length of programs. This correlation does not vary strongly with the plagiarism penalty.

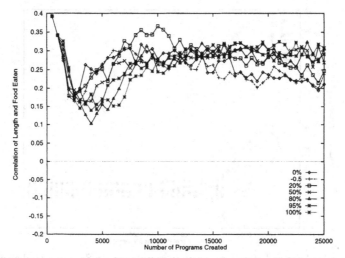

Fig. 6. Correlation of program length and food eaten on Santa Fe trail as plagiarism penalty is increased. Means of 10 runs.

4.2 Fraction of Bloated Programs Used

Every program terminal uses one time unit each time it is executed. This gives us a convenient, if crude, way to estimate the amount of code being used within each program. Figure 7 plots the average number of terminals executed each time a program is run. Initially on average 6.8 terminals are used per program execution, this then rises rapidly to 19.1 before falling back to 9.3 (no penalty). Runs with high penalties are similar except they rise to slightly higher peaks, 26.3, and take longer to fall back to similar values (10.7, 100% penalty). The initial rise in the average may simply be due to selection acting to remove those with lower values, however it appears crossover finds better solutions which execute only a small fraction on their terminals each time the programs are run.

Assume that there are two frequently occurring cases, either the program starts with the ant facing food or facing an empty square. Using this we can estimate the number of terminals contained in the two parts of the program most often used as no more than twice the average number executed each time it is run. I.e. less than 20 by the end of the run. This is clearly a very small fraction of the whole (on average programs exceed 300 nodes, and so must contain at least 150 terminals, by the end of the run, cf. Fig. 3). It appears evolution promotes the creation of small islands of useful code in large programs since such programs are less liable to disruption by crossover. Figure 7 implies the essential nature of the evolved (partial) solutions is unchanged by the plagiarism penalty. Which suggests the penalty reduces the rate at which introns grow but they are still present, even with the highest penalties. This is surprising because the stability of such programs in the face of crossover means they will suffer the penalty which should prevent them creating children in the next generation. We now turn to the dynamic fitness function results, and will see in this respect GP evolves significantly differently.

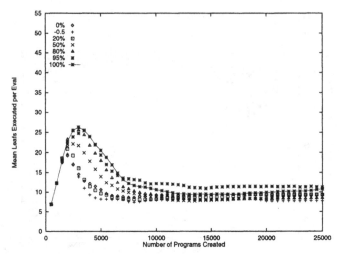

Fig. 7. Time used per call of each program in population on Santa Fe trail as plagiarism penalty is increased. Mean of 10 runs.

5 Random Trails

This section reports results from 50 independent runs with each plagiarism penalty setting. As in Sect. 4 the same initial populations are used with each plagiarism value. Some of the differences between this section and Sect. 4 are due to the larger number or runs made, which reduces the influence of stochastic effects.

To gain a measure of the ability of GP to generalise from the example trials, the populations were also tested on the complete set of 50 trails and statistics for the complete set were gathered. This information is used only for reporting purposes and only the current (and previous) trails influence the course of evolution. (As running all programs on all 50 trails is CPU expensive, this data was only collected for the first ten of the 50 runs in each experiment).

Figure 8 shows the evolution of scores using the new dynamic fitness function. Much of the variation between one generation and the next can be ascribed to the different difficulty of the trials. If we consider Fig. 10 we see rather more monotonic rises in program score. Referring to Fig. 8 again, with 50 program runs, we can see a clear separation into runs with low penalty performing approximately 6–9 food pellets better than those with high penalties. This difference is amplified if instead of looking at average performance, we look at the number of runs where at least one solution to the current trail was found, cf. Fig. 11.

The random trails seem to be a harder problem than the Santa Fe trail, only three runs evolved programs which could follow all 50 trails. Another aspect of this increased difficulty is the evolved solution given in [Koz92, page 154] to the Santa Fe trail scores less than half marks on the random trails. In contrast each of the first programs found which could pass all 50 random trails can also follow the Santa Fe trail.

There is some evidence that the random trails are themselves enough to avoid

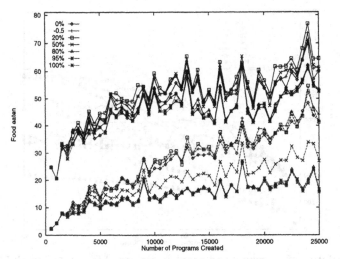

Fig. 8. Evolution of maximum and population mean of food eaten on random trails as plagiarism penalty is increased. Means of 50 runs.

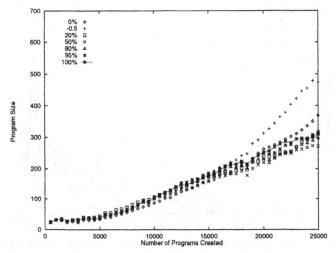

Fig. 9. Evolution of population mean program size on random trails as plagiarism penalty is increased. Means of 50 runs.

convergence of the GP population. Without a penalty there are typically only a handful of programs with the best score by the end of the run. This falls to one or two with large penalties. However lack of convergence would suggest that the fraction of children behaving as their parents would be small, whereas it behaves similarly to the Santa Fe runs and with low penalties (20% or less) the fraction rises to about 80% by the end of the run (cf. Fig. 12). With higher penalties the fraction also behaves like the Santa Fe runs and remains near its initial value until the end of the runs. Again variety is near unity and does not show this convergence at all. (With no penalty then variety reaches 97% on average at generation 50 and with a penalty it is about 95% on average again).

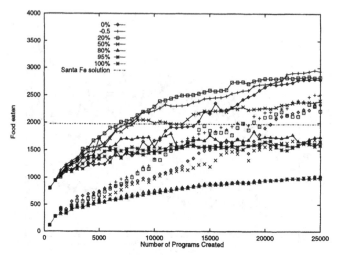

Fig. 10. Evolution of maximum and population mean of food eaten on all 50 trails as plagiarism penalty is increased. Means of 10 runs.

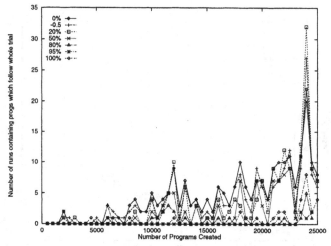

Fig. 11. Number of populations containing a solution to current trail on random trails as plagiarism penalty is increased. 50 runs.

Like the Santa Fe trail, runs with low penalties show convergence in that typically the scores of the parents of the vast majority of the children in the last generation are identical (or have one of two values). With runs with high penalty there is more variation (with on average 29.1 different scores acting as first parents to children in the final population).

5.1 Correlation of Fitness and Program Size

Comparing Fig. 13 with Fig. 5 we see there is no longer a clear separation between runs based upon strength of plagiarism penalty. After generation 15 (7,500

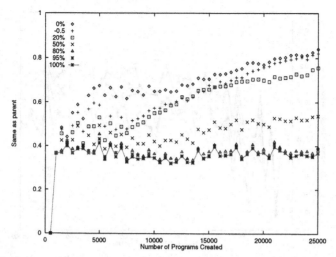

Fig. 12. Evolution of proportion of population with same score as first parent on random trails as plagiarism penalty is increased. Means of 50 runs.

programs created) low penalty runs tend to have smaller covariances than with the Santa Fe trail. This is reflected in the generally lower increase in program size. The lower covariance may be simply due to increased randomness in the fitness function. In contrast when we look at the correlation between score and size (i.e. excluding penalty, cf. Fig. 14) the high penalty runs have an obviously lower correlation. This shows that the penalty has changed the nature of the evolved programs.

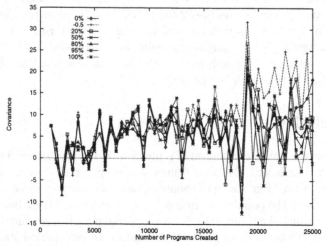

Fig. 13. Evolution of covariance of program length and normalised rank based fitness on random trails as plagiarism penalty is increased. Means of 50 runs.

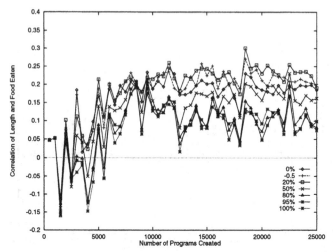

Fig. 14. Correlation of program length and food eaten on random trails as plagiarism penalty is increased. Means of 50 runs.

5.2 Fraction of Bloated Programs Used

Figure 15 confirms in the random trails high plagiarism penalties change the nature of the evolved programs. We see programs with very different behaviour evolve than is the case with the Santa Fe trail or with low penalties. The large number of ant operations per program execution (cf. Fig. 15) might indicate that non-general, or trail specific, programs have been evolved. This is confirmed to some extent if we compare the best scores in the last generation on the last trail with that on all 50 trails. We see performance on the training case (cf. Fig. 8) is proportionately higher than on all 50 trails (cf. Fig. 10). Concentrating on 100% penalty, the mean scores are 52 out of 80, i.e. 66% (averaged over 50 runs) compared to 1590 out of 4,000, i.e. 40% (averaged over 10 runs). That is populations evolved with high plagiarism penalties do considerably better on the immediate training case than they do on the more general one. The difference is not so marked in runs with low penalties.

6 Conclusions

In our earlier work on the evolution of representation size we stressed the importance of individuals with the same fitness as their parents', showing increase in average size in the later stages of evolution could in many cases be ascribed to them dominating the population. In Sect. 4 we introduce a fitness based penalty on programs which don't innovate. Even very large penalties produce only slight reductions in the best of run performance and, in these experiments, cut bloat by about a half.

In the experiments in Sect. 5 we have broaden research into bloat to consider non-static fitness functions. In these experiments a dynamic fitness function also cuts bloat by about a half. We also report combining our dynamic fitness function

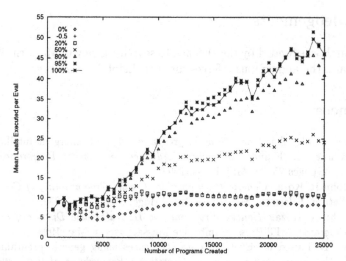

Fig. 15. Time used per call of each program in population on random trails as plagiarism penalty is increased. Mean of 50 runs.

with the plagiarism penalty and note this can also produce a small reduction in the best of run performance but can change the nature of the programs evolved and reduce their ability to generalise.

It is clear that suppressing the large numbers of programs produced in the later stages of conventional GP runs which all have the same performance by using a plagiarism penalty has not prevented bloat completely. In both sets of experiments there is bloating we suspect that this is due to shorter programs in the population being more effected by crossover than longer ones, i.e. their children follow the trails less well.

Our second set of experiments tend to confirm some of the benefits claimed for dynamic fitness measures. E.g. every dynamic fitness run (without a plagiarism penalty) produced programs which performed better on the 50 random trails than the example program evolved on just the Santa Fe trail.

We have deliberately chosen to study ant problems since they are difficult for GP and have properties often ascribed to real world programs (such as rugged landscapes, multiple solutions, competing conventions, poor feedback from partial solutions). Nevertheless it would be interesting to analyse bloat in other GP domains. Further work is needed to understand how GP populations are able to maintain their peak performance even when the selection function appears to prevent direct copying from the best of one generation to the next. This would appear to require constant innovation on the part of the population. Current GP can "run out of steam" so that GP populations stop producing improved solutions [Lan98a, pages 206]. Therefore techniques which encourage constant innovation are potentially very interesting.

Acknowledgements

This research was funded by the Defence Research Agency in Malvern. We would like to thank the anonymous referees for their helpful comments.

References

[Ang94]   Peter John Angeline. Genetic programming and emergent intelligence. In Kenneth E. Kinnear, Jr., editor, *Advances in Genetic Programming*, chapter 4, pages 75–98. MIT Press, 1994.

[Koz92]   John R. Koza. *Genetic Programming: On the Programming of Computers by Natural Selection*. MIT Press, Cambridge, MA, USA, 1992.

[Koz94]   John R. Koza. *Genetic Programming II: Automatic Discovery of Reusable Programs*. MIT Press, Cambridge Massachusetts, May 1994.

[Lan95]   W. B. Langdon. Evolving data structures using genetic programming. In L. Eshelman, editor, *Genetic Algorithms: Proceedings of the Sixth International Conference (ICGA95)*, pages 295–302, 1995. Morgan Kaufmann.

[Lan97]   W. B. Langdon. Fitness causes bloat in variable size representations. Technical Report CSRP-97-14, University of Birmingham, School of Computer Science, 14 May 1997. Position paper at the Workshop on Evolutionary Computation with Variable Size Representation at ICGA-97.

[Lan98a] W. B. Langdon. *Data Structures and Genetic Programming*. Kulwer, 1998.

[Lan98b] W. B. Langdon. The evolution of size in variable length representations. In *1998 IEEE International Conference on Evolutionary Computation*, Anchorage, Alaska, USA, 5-9 May 1998. Forthcomming.

[LP97]   W. B. Langdon and R. Poli. Fitness causes bloat. In P. K. Chawdhry *et al*, editors, *Second On-line World Conference on Soft Computing in Engineering Design and Manufacturing*. Springer-Verlag London, 23-27 June 1997.

[LP98a]   W. B. Langdon and R. Poli. Fitness causes bloat: Mutation. This volume.

[LP98b]   W. B. Langdon and R. Poli. Why ants are hard. Technical Report CSRP-98-4, University of Birmingham, School of Computer Science, January 1998.

[NB95]   Peter Nordin and Wolfgang Banzhaf. Complexity compression and evolution. In L. Eshelman, editor, *Genetic Algorithms: Proceedings of the Sixth International Conference (ICGA95)*, pages 310–317, 1995. Morgan Kaufmann.

[Pri70]   George R. Price. Selection and covariance. *Nature*, 227, August 1:520–521, 1970.

[SFD96]   Terence Soule, James A. Foster, and John Dickinson. Code growth in genetic programming. In John R. Koza, *et al*, editors, *Genetic Programming 1996: Proceedings of the First Annual Conference*, pages 215–223, 1996. MIT Press.

[Tac93]   Walter Alden Tackett. Genetic programming for feature discovery and image discrimination. In Stephanie Forrest, editor, *Proceedings of the 5th International Conference on Genetic Algorithms, ICGA-93*, pages 303–309, University of Illinois at Urbana-Champaign, 17-21 July 1993. Morgan Kaufmann.

[Tac94]   Walter Alden Tackett. *Recombination, Selection, and the Genetic Construction of Computer Programs*. PhD thesis, University of Southern California, Department of Electrical Engineering Systems, 1994.

[Tac95]   Walter Alden Tackett. Greedy recombination and genetic search on the space of computer programs. In L. D. Whitley and M. D. Vose, editors, *Foundations of Genetic Algorithms 3*, pages 271–297, 1995. Morgan Kaufmann.

Speech Sound Discrimination with Genetic Programming

Markus Conrads[1] Peter Nordin[12] and Wolfgang Banzhaf[1]

[1] Dept. of Computer Science, University of Dortmund, Dortmund, Germany
[2] Dacapo AB, Gothenburg, Sweden

Abstract. The question that we investigate in this paper is, whether it is possible for Genetic Programming to extract certain regularities from raw time series data of human speech. We examine whether a genetic programming algorithm can find programs that are able to discriminate certain spoken vowels and consonants. We present evidence that this can indeed be achieved with a surprisingly simple approach that does not need preprocessing. The data we have collected on the system's behavior show that even speaker-independent discrimination is possible with GP.

1 Introduction

Human speech is very resistant to machine learning approaches. The underlying nature of the time-dependent signal with its variable features in many dimensions at the same time, together with the intricate apparatus that has evolved in human speech production and recognition make it one of the most challenging – and at the same time most rewarding – tasks in machine pattern recognition.

The strength of Genetic Programming [Koz92] is its ability to abstract an underlying principle from a finite set of fitness cases. This principle can be considered the essence of the regularities that determine the appearance of concrete fitness cases. Genetic programming tries to extract these regularities from the fitness cases in the form of an algorithm or a computer program. The particular implementation of genetic programming that we use enables us to evolve computer programs at the very lowest level, the machine code level.

Despite their potential complexity it is obvious that regularities rule human speech production, otherwise oral communication would never be possible. These regularities, however, must already reside in the "raw" time signal, i.e. the signal void of any preprocessing, because any method of feature-extraction can only extract what is already there.

The question, then, that we investigate in this paper is whether it it possible for a genetic programming system to extract the underlying regularities of speech data from the raw time signal, without any further preprocessing or transformation.

2 Classification Based on Short-Time Windows

2.1 The Goal

At an abstract level, the task of the genetic programming system is to evolve programs based on the time signal that are able to distinguish certain sounds. Since the task is complicated by the fact that at reasonable sampling rates there is a huge amount of data, it is necessary to impose certain restrictions.

For classification we shall use a short window for the time signal. The size of this window will be 20 ms. At those window sizes the (transient) speech signal can be regarded as approximately stationary. Thus classification of more complex transient signals from longer time periods will not be studied here.

Some of the challenges with our approach will be summarized in the sections to follow.

2.2 Size of the Data Set

When using the time signal as a basis for classification, the system will have to cope with very large data sets. The samples used here have been recorded with a sampling rate of 8 kHz, which means that for each 20 ms window of the signal we have 160 sampled values.

It would be very hard, if not outright impossible, to evolve a GP-program with 160 independent input variables. There would be far too many degrees of freedom, which could easily lead to over-fitting. Besides, the programs would have to become very complex: Reading such a large input and filtering out distinctive features for classification requires a very large amount of instructions.

In many automatic speech recognition systems of today this problem is solved by calculating a feature-vector. Applying a Fourier transform, a filter bank, or other feature extraction measures does not only emphasize the important features, it also drastically reduces the amount of data to be fed into the learning system.

In this study, however, we will not allow any of these measures, as we want to show that genetic programming is able to find discriminating features in the time signal *without* any preprocessing applied to it. We must find a way, though, to reduce the input to the GP-program without reducing or changing the data.

2.3 Scanning the Time Signal

The solution to the problem of the previous section we embarked on is the following. If it is not possible to give the program all data at once, we shall feed them step by step: The time signal is scanned from start to end. The GP program is expected to run through an iteration where in every iteration step it gets only *one* sampled value plus the current position of the sampled value. The program then produces an *output vector* which consists of different components. Additionally, in every iteration step the program gets the output vector of the

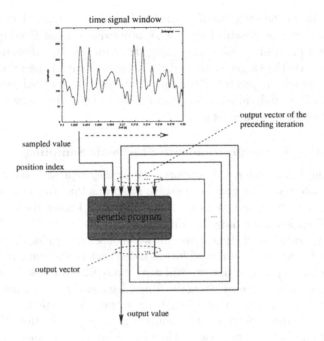

Fig. 1. Data flow when classifying a time signal window

preceding iteration. For the first iteration, we initialize this vector to 0 in each component. Figure 1 illustrates our approach.

The feedback from the output is important. It makes the behavior of the program in iteration n dependent not only on the current input but also on the behavior in the previous step, $n - 1$, which again depends on what happened in the step before and so on. With this method, the output produced in step n depends on *all* sampled values read up to that moment.

Thus, the output of the last iteration is influenced by all sampled values in the scanned time signal window. This output can be used for classification of the signal. For this classification we use only one component of the output vector, leaving it up to the system to use the others for itself on its way to solve the problem.

Relying on such an approach dramatically reduced the number of inputs to the program from a three-digit number to a one-digit number, without applying any preprocessing to the data.

2.4 Calculating the Fitness

In the present study, we only examine classification into two classes. Experiments using more than two classes have been very discouraging so far and will not be considered here.

In order to distinguish two classes $\mathcal{A}$ and $\mathcal{B}$ we determine a target value for each class, $T_{\mathcal{A}}$ and $T_{\mathcal{B}}$. Here we use the values 10,000 and 0. This means that if the output value of the program after the last iteration is bigger than 5,000, the scanned sample will be classified as class $\mathcal{A}$, otherwise as class $\mathcal{B}$ sample.

Fitness of a program is calculated using the *error*, i.e. the absolute value of difference between the target value and the actual output of the program. The sum of all errors for the presented training instances, the so called *fitness cases*, is the fitness of the individual. Thus, the fitness value is the lower the better, the best possible fitness value is 0.

2.5 Improving Generalization with Stochastic Sampling

When choosing fitness cases, it is important that they represent the problem in the best possible way. The goal is to evolve programs that can *generalize*. It is not very useful to have programs that work good on known data, the fitness cases, but fail on unknown data (overfitting).

In order to avoid overfitting, how can we choose the "right" fitness cases? We will have to choose a sample of fitness cases that is adequate to the problem. This means in practice that we will need many fitness cases. Using a large amount of fitness cases for evaluating the fitness has, on the other hand, a serious disadvantage: It will require a very long time to train the system.

There is a method, however, to maintain good generalization abilities by using a large number of fitness cases while not requiring that much computing power, *stochastic sampling*[1] [NB95c,NB97]: From a large set of fitness cases a small sample is chosen randomly before a fitness evaluation takes place. Fitness evaluation is performed only on the small sample, and hence is less expensive than evaluation on the entire fitness case set.

Thus, no program from the population can really "know" the data it will be confronted with. There are not many fitness cases when fitness is calculated, and memorizing does not help, because at the next evaluation a completely different set is presented. Thus, programs must develop a strategy for how to cope with many *different* possible inputs. They are forced to really *learn* instead of just to memorize. Learning in this context means to perform an *abstraction process*.

Of course, at any moment a relatively good program may be in "bad luck" and be competing against a program that might work worse on most other fitness cases. But seen over many tournaments in the evolutionary process, only those programs will survive that have developed a strategy to cope with more fitness cases than other programs.

A further advantage of this method is that the total amount of fitness cases can be increased arbitrarily, without slowing down the system because the number of fitness cases used for each fitness evaluation remains the same. It can be expected, that a larger set of fitness cases will result in better generalization because the generated GP programs will have to be able to cope with an even larger set of data.

Reports in the literature [GR94] on the failure of this stochastic selection method for training do not conform to our experiences.

---

[1] A similar method called *Random Subset Selection*, which uses a slightly different way of selecting the fitness cases, has been described in [GR94].

j fitness-case no.	C_{in} input-sample	C_{out} desired output
0	sample of an [a]	10000
1	sample of an [a]	10000
$\vdots$	$\vdots$	$\vdots$
$M/2-1$	sample of an [a]	10000
$M/2$	sample of an [i]	0
$M/2+1$	sample of an [i]	0
$\vdots$	$\vdots$	$\vdots$
$M-1$	sample of an [i]	0

Table 1. Template of a fitness case table used to evolve a program that can distinguish between [a] and [i]. A simple measure to help the system succeed in the classification task is to provide approximately the same amount of fitness cases for each class.

2.6 Applying Stochastic Sampling to Speech Sound Discrimination

The classifiers that we want to evolve will work on a short time window. The speech sounds that the system will need to classify are vowels, fricatives or nasals. These are sounds which can be articulated for a longer time than 20 ms and hence can be regarded as approximately stationary.

If we record a speaker producing an [a] for half a second and sample the signal with a frequency of 8 kHz, we get 4000 sampled values. Therefore we might put the 20 ms window on the time signal at 4000 different positions. Because the signal always changes at least slightly over time, we hence gain 4000 different fitness cases out of one half second of recording. The large amount of possible fitness cases is ideal for stochastic sampling: Calculating all fitness cases would be far too expensive, but their large number will be very helpful for avoiding overfitting. Before fitness of an individual program with respect to a certain sample of fitness cases is evaluated, a window is layed upon the sample at a random position.

In order to ensure that programs do not over-adapt to a certain act of speaking, fitness should be calculated using samples of different acts of speaking. It is advisable to establish a fitness case table for registering which samples are being used and which class they belong to, i.e. what the desired output should be. Table 1 shows a fitness case table used to evolve an "[a]-[i]-classifier" program that can distinguish between [a] and [i]. Fitness case tables for evolving other classifiers can be constructed in similar ways. It is important to note, that the sounds to be classified do not contain significant time-structured features. For instance, for classifying sounds like diphthongs or plausives, the method would have to be modified.

If instead of one-speaker samples those of many speaker are used for the training, it is also possible to reach speaker-independent phone recognition[2]. Samples recorded from many different speakers enable the system to abstract from speaking habits of a single or a few speakers.

3 A Run Example

In this section we shall demonstrate the dynamics of our system with an example run. The phenomena documented here have been observed in many other runs, so the documentation should be considered representative. The behavior of the system applied to other phone discrimination problems is qualitatively similar, though different recognition rates and fitness values will occur in each case.

3.1 The GP-System and Its Parameters

We used the *Automatic Induction of Machine Code by Genetic Programming* (AIMGP) system[3] for this problem. AIMGP is a linear GP system that directly manipulates instructions in the machine code. We call it a "linear" system because the representation of the evolved programs is instruction-after-instruction, and therefore linear as opposed to the hierarchical, tree-based structure in most common GP systems.[4] Theoretically, any GP system could be used for this problem, and there are already many [BNKF98]. The problem addressed here, however, is quite complex and requires a lot of computing power. The advantage of AIMGP compared to other GP systems is its speed (at least a factor of ten compared to other existing GP systems) and it is therefore better suited for the challenge of this task. In addition, the representation of individuals is very efficient in AIMGP, giving us the opportunity to run large populations (10,000 individuals) on a single processor system (SPARC), which turned out to be necessary to solve this problem [Nor97].

The parameter settings are shown in Table 2. The choices shown have generated good results, although a parameter optimization was not done. Our goal was not to succeed with parameter tweaking, but in principle.

3.2 The Fitness Case Table

In our example run we want to evolve a program that can distinguish between [a] and [i]. The fitness case table consists of two samples of each vowel, spoken by the same speaker (see Table 3). So, the system is set to develop a speaker-dependent classifier.

---

[2] Phones are subunits of phonemes.

[3] formerly known as CGPS [Nor94,NB95b]

[4] The linear representation seems to have merits of its own, not only regarding the speedup due to directly executing machine code instructions. [Nor97]

Parameter	Setting
population size:	10 000
selection:	tournament
tournament size:	4 (2 x 2)
maximum number of tournaments:	5 000 000
mutation-frequency (terminals):	20 %
mutation-frequency (instructions):	35 %
crossover-frequency:	100 %
maximum program size:	256 instructions
maximum initial size:	30 instructions
maximum initial value of terminal numbers:	100
number of input-registers:	3
number of output(memory)-registers:	1
number of ADFs:	0
instruction set:	addition subtraction multiplication left shift right shift bitwise logical OR bitwise logical AND bitwise logical XOR
termination-criterion:	exceeding the maximum number of tournaments

Table 2. The Koza tableau of parameter settings for AIMGP

Sample	Target Value
sample 1 of an [a]	10000
sample 2 of an [a]	10000
sample 1 of an [i]	0
sample 2 of an [i]	0

Table 3. Fitness case table for developing a program that can distinguish between [a] and [i]. All recordings have been taken from the same speaker.

3.3 Fitness Development

In Figure 2 the fitness development is shown for evolving an [a]-[i]-classifier. Due to the fitness value being calculated as the sum of errors (see section 2.4), the most interesting fitness value is 20,000. We call it the *critical value*. Caused by the structure of the fitness cases, this value can be reached by programs that have a constant output of 5,000, no matter what the input might be. This trivial "solution" leads to a recognition-rate of 50% in a two-class discrimination problem.

As soon as the mean fitness falls below the critical value the ability for discriminating the two classes is indicated. For a program to fall below the critical value is far from trivial. The critical value is a local optimum, because other programs that produce a non-constant output might easily be wrong and go extinct. For this reason it is important to set the population size large enough, so that some of the programs with non-constant output are still able to survive. These programs might eventually breed later on and escape the local optimum.

Therefore, the progression of the mean fitness during the first 1,000,000 tournaments in Figure 2 is particularly interesting. One can see it falling very fast to the critical value, but not falling below. For about 400,000 tournaments the mean fitness stagnates at the critical value. Afterwards it suddenly increases again, reaches its maximum at 600,000 tournaments and then descends, falling below the critical value at about 850,000 tournaments. Fitness keeps decreasing, until it reaches a value of about 5,000.

How can this progression be explained? At first, the local optimum is quickly found. Then, for about 400,000 tournaments, nothing seems to happen if we observe mean fitness only. But if we observe fitness of the best program, it becomes apparent that something is going on indeed. Best fitness is considerably lower than mean fitness and heavy fluctuations are observable. Recall that the system works with stochastic sampling. The same individual may have very different fitness-values, depending on which fitness case is chosen. Especially in the first phase of the evolutionary process this can lead to heavy fluctuations in fitness when observing a special individual. Note also, that the best individual is not always the same, so the fact that it consistently remains below the critical value (from the outset) does not mean that one individual has already found a solution.

But why is it that the mean fitness increases again sharply? Is there something wrong with the process, is the population getting worse? No, just the opposite. A raise in average fitness indicates that the system is about to discover or already has discovered a solution. This solution, however, is not very robust. It can be destroyed by crossover or mutation events, and also due to bad-luck resulting from stochastic sampling. At this point, a look at the best fitness reveals that it is not changing that much anymore during this period, giving an additional hint that a solution has been found already. As evolution continues, the individuals that have found the discriminating features are busy breeding and becoming more robust. They will finally take over the whole genetic pool.

3.4 Development of Program Length

In Figure 2 we depict the length development of programs over the same time period as that of Figure 2. One can see immediately that length increases dramatically at about 1 million tournaments. The limit for program length is set to 256. This limit is reached by both the average program length and the length of the best individual after about 2 million tournaments.

The interesting point about this dynamics is that the dramatic grow of program length happens at the time when fitness decreases. This means that long

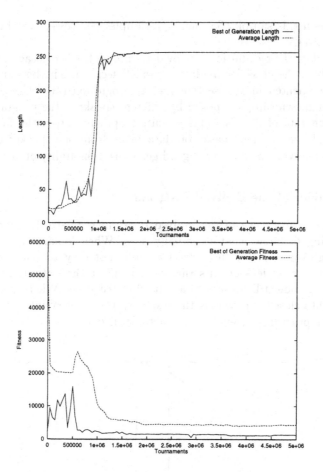

Fig. 2. The development of the program lengths (above) compared with the fitness development (below) when evolving a [a]-[i]-classifier. The x-axis gives the amount of tournaments performed. One tournament is equivalent to four fitness evaluations, 5000 tournaments are equivalent to one generation.

programs have an advantage compared to short ones. We might want to conclude that longer programs are more resistant against destructive mutation or crossover as we have pointed out elsewhere [NB95a,BNKF98] in more details.

3.5 Performance of the Evolved Individual

As a measure of the performance of the evolved individual we calculate the classification rate: Before the run, ten samples of each vowel had been recorded by the same speaker. Two samples have been used for evolving the classifier already. The eight samples left are unknown to the system. These samples are now used to calculate the classification rate of the best-of-run program. Each sample is about 0.5 s long which means that we can lay about 4000 time windows of 20 ms upon this sample. So, having 8 samples per vowel, we obtain 64 000 *validation cases.*

We calculate the recognition rate by determining the percentage of validation cases on which the classifier made the correct decision. The best individual of the run documented in this section had a recognition rate of 98.8 %. When tested on unknown samples spoken by different speakers, the program still had a recognition rate of 93.7 %. This is quite surprising, given the fact that the program had been evolved based on data taken from one speaker only. Thus, there is strong evidence for a very good generalization ability of our system.

3.6 Behavior of the Evolved Program

Because the performance of the evolved program is relatively good, it would be interesting to know what it actually does. When looking at the program code, though, one quickly discovers that it is not easy to understand. Even after introns, i.e. code segments that do not affect the output, are removed, the remaining code still consists of about 60 instructions. When the program is converted into closed expressions the result is also not very intuitive. After all, GP develops programs by evolution, not by intuition.

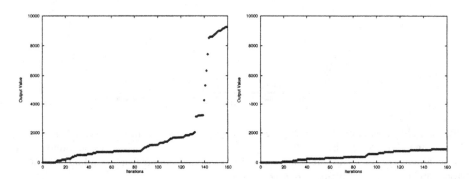

Fig. 3. The dynamics of the output value of an [a]-[i]-classifier working on a sample of an [a] (left side) resp. an [i] (right side).

We therefore approach this problem empirically by studying program behavior. Figure 3 shows the dynamics of the output-value when the classifier scans an [a] and an [i], respectively. Each dot represents the output-value after one of the 160 necessary calls during scanning a 20 ms time window. For an [a], the target-value is 10 000 (case A), for an [i] it is 0 (case B).

In both figures one can see a monotone increase of the output value. However, there is a sudden step in case A from 2000 up to 8000. After this step the value is increasing steeper than before. In case B no such step can be observed.

This transition occurred in all evolved classifiers we have investigated. Sometimes there was even more than one step. The time of the transition is not the same for all input data. The transition may be delayed, depending on the position of the time-window. Probably this avoids a dependence on a phase-lag caused by the position of the time-window.

Some hints how the program might work can be obtained from *linear digital filters*. These filters work in a similar fashion as the evolved classifiers. The section 3.8 will discuss this topic in more detail.

3.7 An Undiscovered Bug

When looking at the closed expressions which evolved, we made a very astonishing discovery: Only one of the memory-registers (originally 6), intended to be used for feedback of the output (see section 2.3), was employed by the system! The only memory register used was the one that simultaneously served as single output register. That register was used subsequently for calculating the fitness of the individual program, after all 160 samples had passed through it.

Unfortunately, the reason for this parsimonious behavior of programs was a bug in our system! All memory registers (except the one also serving as output) were erased or at least changed in some way between calls and were therefore of no use in feedback.

This bug seems amazingly severe. Instead of the seven potentially independent feedback signals that we intended to provide originally, the system had only one. Although the results presented here have been produced by this unintentionally weak system, they are pretty good. The system had learned to cope with this weakness and did not use the additional registers in feedback, for their lack of stability. Instead, they were put to use as additional calculation registers for storage of provisional results *within* the program.

This whole episode can be seen as another example for the robustness of the GP paradigm, which has been observed by other researchers before. Kinnear notes on this topic in [Kin94]:

> "Few of us have experience with truly robust computer algorithms or systems. Genetic Programming is such a system, and through the same power that allows it to search out every loophole in a fitness test it can also sometimes manage to evolve correct solutions to problems despite amazingly severe bugs."

3.8 Linear Digital Filters and Evolved Classifiers

How Do Linear Digital Filters Work? Digital filters work by extracting certain frequencies from the time signal: A linear digital filter computes a sequence of output-points y_n from a sequence of input-points x_k, using the formula (from [PFTV88]):

$$y_n = \sum_{k=0}^{M} c_k x_{n-k} + \sum_{j=1}^{N} d_j y_{n-j} \tag{1}$$

The $M + 1$ coefficients c_k and the N coefficients d_j are fixed in advance and will determine the filter response. The filter produces a new output using the current input and the M earlier input-values before, and also using the N outputs produced earlier. If N is 0 the filter is called non-recursive or *finite impulse response* (FIR)-filter and the former output does not affect the current output. If $N \neq 0$ the filter is called recursive or *infinite impulse response* (IIR)-filter.[5]

What is the use of these filters? They are used to extract certain frequencies in a signal and to dampen others. This behavior is expressed by the *filter response function*. For linear digital filters it reads [PFTV88]:

$$\mathcal{H}(f) = \frac{\sum_{k=0}^{M} c_k e^{-2\pi i k(f\Delta)}}{1 - \sum_{j=1}^{N} d_j e^{-2\pi i j(f\Delta)}} \tag{2}$$

With this formula one can calculate the filter response if the c and d coefficients are given. In order to *design* a filter, however, one must go the other way: The c's and d's are to be determined for a desired filter-response. For further information on filter design, the reader is referred to [HP91,MK93].

Are The Evolved Programs Digital Filters? When looking at the algorithm for digital filters, they appear similar to our GP-approach for classifying speech data. In an IIR-filter there is a *feedback* of the output. The N formerly produced outputs are being used to produce the current output. The data-flow is very similar to the one shown in Figure 1. There, too, several output-values are being fed back, although the output-values are all from the same iteration.

The GP system, however, certainly does not evolve a linear response. Instead, it uses bit- and shift-operations plus multiplication operations all of which result in non-linear behavior. Irrespective of the nature of the filter, formula (1) suggests that GP-evolved programs might be able to extract features from a time signal in quite a similar way. Although linear digital filters provide a hint, the resulting programs are much more complicated than a simple linear digital filter.

4   Experimental Results

4.1   The Quality of the Testing-Data

For the experiments documented here sound-samples of several speakers have been recorded. The recording was done with the standard sound-device of a SPARCstation 10. Samples are in 8 kHz μ-law format. They have been converted to a linear 8 bit format. Except for this conversion the data had not been changed in any way. The recordings are not free from noise. Although there are no background voices on the recordings, the fan is audible to the human ear.

---

[5] Infinite impulse response means, that a filter of this kind may have an response that extends indefinitely (due to the feedback of the output). In contrast, finite impulse response filters have a definite end to their response if there is no input-signal anymore.

In summary, the quality of the recordings is not very high. With a sampling rate of 8 kHz only signal frequencies below 4 kHz can be played back reliably. For certain speech sounds this is definitely not enough. One cannot, for instance, distinguish between [s] and [f] when talking on the phone, a transmission of comparable quality.

4.2 Speaker-Dependent Vowel Discrimination

For vowel discrimination we used the vowels [a], [e], [i], [o] and [u]. For each pair of vowels we evolved classifiers, all in all ten different types of classifiers. In the speaker-dependent case all recordings used to train the system and to test its performance were spoken by the same person.

For determining the recognition rate we used the eight unknown samples of each of the two vowels the classifier is able to distinguish. Each classifier scans a time window of 20 ms. This window is shifted over the entire sample. In each iteration the window starts 0.12 ms later, which corresponds to exactly one sampled value at a sampling rate of 8 kHz. The number of hits, i.e. correct classification answers, is counted and a percentage is computed. Two samples of different classes form a pair. The hit-rate of a pair is calculated as the mean of the hit-rates of each sample in the pair. The longer sample is cut off in order to have an equal number of chances to classify correctly.

For each classifier we get eight pairs. The classification rate is computed as the mean of the hit-rates of all eight pairs. Table 4 shows the recognition rates for all possible vowel combinations. For each vowel combination five GP runs have been done. Table 4 shows the best value that could be reached in these runs.

	[a]	[e]	[i]	[o]	[u]
[a]	—	93.1	99.1	75.0	50.0
[e]	—	—	85.2	74.0	79.1
[i]	—	—	—	89.4	76.1
[o]	—	—	—	—	75.4
[u]	—	—	—	—	—

Table 4. The recognition rates for vowel classification in the speaker-dependent case (in %)

Looking at these values it becomes apparent, that there are obviously some combinations that make it easier to classify than others. These combinations are [a] and [e] and also [a] and [i], where recognition rates of more than 90 % could be obtained (for [a] and [i] we even obtained over 99 %, which can be considered perfect). For all other combinations we have recognition rates of over 70 %. An exception is the combination [a] and [u], where no solution could be found.

We think these results may even be improved by doing more runs or optimizing GP run parameters.

4.3 Speaker-Independent Vowel Discrimination

As said earlier, in order to evolve a classificator that can discriminate vowels irrespective of the speaker, it is necessary to have training data spoken by as many people as possible. For this test series we had training and testing data spoken by six persons. For reliable speaker-independence this is, of course not enough diversity. However, our aim was to show that speaker independence is possible at all. With six different speakers, the system will already have to cope with a lot of in-class-variance.

The recognition rates were determined in the same way as before. The only difference was that training and testing samples have been recorded from six *different* speakers.

Table 5 shows the best recognition rates that could be obtained. It is somewhat surprising that the combinations with best results are not always the same as in the speaker-dependent case. [a] and [o], as well as [a] and [u] could be discriminated very well in this case although this was not possible in the speaker-dependent case. On the other hand results for [a] and [e] as well as [e] and [i] are worse.

	[a]	[e]	[i]	[o]	[u]
[a]	—	72.2	92.2	88.6	96.4
[e]	—	—	78.9	61.3	69.7
[i]	—	—	—	81.1	65.2
[o]	—	—	—	—	60.0
[u]	—	—	—	—	—

Table 5. The recognition rates for vowel discrimination in the speaker-independent case (in %)

Speaker-independent vowel discrimination is more difficult, because the in-class variance is much higher. Therefore, probably more runs would have been necessary to reach the same level of recognition rates. However, it is clear already from these limited results that distinctive features could be found by GP that were exploited for discrimination.

4.4 Discriminating Voiced and Unvoiced Fricatives

In the third task we examined, discrimination between voiced and unvoiced fricatives there are now more than one phones in each class. In this test series there are three each; the voiced fricatives [z], [ʒ] and [v], and their unvoiced pendants [s], [ʃ] and [f]. Three phones in each class mean a still larger in-class variance than before.

The recognition rates can be read from Table 6. One can see that the value for all combinations with [ʒ] is at about 60 %, while the rates are much higher in all other combinations. The reason could be that [ʒ] differs too much from [z] and [v]. Recall that the recordings have been sampled with 8 kHz, a quality that makes it difficult even for humans to distinguish between certain sounds. Signals that contain high frequencies loose information and fricatives have high frequency formants.

	[z]	[ʒ]	[v]
[s]	90.4	59.6	81.6
[ʃ]	90.6	59.5	81.2
[f]	90.5	60.0	81.3

Table 6. The recognition rates classification of voiced and unvoiced fricatives (in %). Only one classifier used.

In the other cases the GP system has found a way to separate the classes. Recognition rates of 80 to 90 % could be obtained.

4.5 Discriminating Nasal and Oral Sounds

Like in the discrimination task of voiced versus unvoiced sounds this problem has several sounds in one class. In our test series we evolved programs that can discriminate the nasal sounds [n] and [m] from the vowels [a] and [e]. This is another example of large in-class variance, since in the sections 4.2 and 4.3 the vowels [a] and [e] were *discriminated*, while here they have to fall into the same class.

Table 7 shows our results. Again, the system was able to find distinctive features that can be applied to all different phones in the two classes.

	[a]	[e]
[n]	89.3	78.8
[m]	97.0	86.4

Table 7. The recognition rates for classification of oral versus nasal sounds (in %). Only one classifier used.

5 Conclusions

Looking at our experimental results, the most important conclusion is that in every case it was possible for the GP system to evolve programs able to filter out certain discriminating features from the time signal. Even if in some cases the recognition rates were not optimal, there was always a tendency toward discrimination.

Thus, the question that we started with; Is it possible for genetic programming to extract an underlying law from the raw time signal in order to make a classification? can be positively answered. Indeed, the system has extracted regularities, regularities at least, that underlie a large part of the fitness cases, and that are helpful for discriminating the two classes.

In the speech recognition literature, it is often stated that the time signal would not be suitable for classification. It would not contain the needed features in an easily accessible way. When looking at the time signal one can indeed be amazed, how similar signals of different phones can look in some cases, and how different signals of the same phone can look in another time window. But this is clearly no limitation to the GP system.

What is particularly interesting in the GP system is the way, in which it classifies. Most methods used in speech recognition (or more general pattern recognition) perform some kind of comparison of a feature-vector with some reference-vector. Nothing like that is happening here. The evolved programs are (after removing the introns) much too short to memorize reference vectors. They *have to* use the regularities residing in the data to discriminate classes.

Another interesting aspect of this approach is, that the evolved classifiers are very fast, for various reasons:

- No additional preprocessing is needed;
- The programs are already represented in machine code;
- Introns[6] (code that does not effect the output) can be removed from the final program, resulting in a speedup of approximately a factor of 5.

It is certainly too early to say how far this approach might lead, but we see at least some potential for applications to be developed along these lines. The evolution of machine code results in very rapidly executing code that can - by virtue of being the native language of the CPU - perform well even on weak platforms.

---

[6] In chapter 7 of [BNKF98] further information on the emergence of introns in GP can be found.

ACKNOWLEDGMENT

Support has been provided by the DFG (Deutsche Forschungsgemeinschaft), under grant Ba 1042/5-1.

References

[BNKF98] Wolfgang Banzhaf, Peter Nordin, Robert Keller, and Frank D. Francone. *Genetic Programming — An Introduction.* dpunkt/Morgan Kaufmann, Heidelberg/San Francisco, 1998.

[GR94] Chris Gathercole and Peter Ross. Some training subset selection for supervised learning in genetic programming. In Yuval Davidor, Hans-Paul Schwefel, and Reinhard Männer, editors, *Parallel Problem Solving from Nature III*, volume 866 of *Lecture Notes in Computer Science*, pages 312–321, Jerusalem, 9-14 October 1994. Springer-Verlag, Berlin, Germany.

[HP91] Richard A. Haddad and Thomas W. Parsons. *Digital signal processing. Theory, applications, and hardware.* Electrical engineering, communications, and signal processing series. W H Freeman, New York, NY, 1991.

[Kin94] K. E. Kinnear, editor. *Advances in Genetic Programming.* MIT Press, Cambridge, MA, 1994.

[Koz92] John R. Koza. *Genetic Programming – On the Programming of Computers by Means of Natural Selection.* MIT Press, Cambridge, MA, 1992.

[MK93] Sanjit K. Mitra and James F. Kaiser, editors. *Handbook for digital signal processing.* A Wiley-Interscience publication. Wiley, New York, 1993.

[NB95a] Peter Nordin and Wolfgang Banzhaf. Complexity compression and evolution. In L. Eshelman, editor, *Genetic Algorithms: Proceedings of the Sixth International Conference (ICGA95)*, pages 310–317, Pittsburgh, PA, 15-19 July 1995. Morgan Kaufmann, San Francisco, CA.

[NB95b] Peter Nordin and Wolfgang Banzhaf. Evolving Turing-complete programs for a register machine with self-modifying code. In L. Eshelman, editor, *Genetic Algorithms: Proceedings of the Sixth International Conference (ICGA95)*, pages 318–325, Pittsburgh, PA, 15-19 July 1995. Morgan Kaufmann, San Francisco, CA.

[NB95c] Peter Nordin and Wolfgang Banzhaf. Genetic programming controlling a miniature robot. In E. V. Siegel and J. R. Koza, editors, *Working Notes for the AAAI Symposium on Genetic Programming*, pages 61–67, MIT, Cambridge, MA, 10–12 November 1995. AAAI, Menlo Park, CA.

[NB97] Peter Nordin and Wolfgang Banzhaf. An on-line method to evolve behavior and to control a miniature robot in real time with genetic programming. *Adaptive Behavior*, 26:107 – 140, 1997.

[Nor94] Peter Nordin. A compiling genetic programming system that directly manipulates the machine code. In Kenneth E. Kinnear, Jr., editor, *Advances in Genetic Programming*, chapter 14, pages 311–331. MIT Press, Cambridge, MA., 1994.

[Nor97] Peter Nordin. *Evolutionary Program Induction of Binary Machine Code and its Applications.* PhD thesis, University of Dortmund, M"unster, 1997.

[PFTV88] William H. Press, Brian P. Flannery, Saul A. Teukolsky, and William T. Vetterling. *Numerical Recipes in C - The Art of Scientific Computing.* Cambridge University Press, Cambridge, 1988.

Efficient Evolution of Asymmetric Recurrent Neural Networks Using a PDGP-inspired Two-Dimensional Representation

João Carlos Figueira Pujol and Riccardo Poli

School of Computer Science
The University of Birmingham
Birmingham B15 2TT, UK
E-MAIL: {J.Pujol,R.Poli}@cs.bham.ac.uk

Abstract. Recurrent neural networks are particularly useful for processing time sequences and simulating dynamical systems. However, methods for building recurrent architectures have been hindered by the fact that available training algorithms are considerably more complex than those for feedforward networks. In this paper, we present a new method to build recurrent neural networks based on evolutionary computation, which combines a linear chromosome with a two-dimensional representation inspired by Parallel Distributed Genetic Programming (a form of genetic programming for the evolution of graph-like programs) to evolve the architecture and the weights simultaneously. Our method can evolve general asymmetric recurrent architectures as well as specialized recurrent architectures. This paper describes the method and reports on results of its application.

1 Introduction

The ability to store temporal information makes recurrent neural networks (RNNs) ideal for time sequence processing and dynamical sytems simulation. However, building RNNs is far more difficult than building feedforward neural networks. Constructive and destructive algorithms, which combine training with structural modifications which change the complexity of the network, have been proposed for the design of recurrent networks [1, 2], but their application has been hindered by the fact that training algorithms for recurrent architectures are considerably more complex than their feedforward counterparts [3, 4, 5, 6, 7].

Recently, new promising approaches based on evolutionary algorithms, such as evolutionary programming (EP) [8] and genetic algorithms (GAs) [9], have been applied to the development of artificial neural networks (ANNs). Approaches based on EP operate on the neural network directly, and rely exclusively on mutation [10, 11, 12, 13] or combine mutation with training [14]. Methods based on genetic algorithms usually represent the structure and the weights of ANNs as a string of bits or as a combination of bits, integers and real numbers [15, 16, 17, 18, 19, 20], and perform the crossover operation as if the network were a linear structure. However, neural networks cannot naturally be represented as vectors. They are oriented graphs, whose nodes are neurons and whose arcs are synaptic connections. Therefore, it is arguable that any efficient approach to evolve ANNs should use operators based on this structure.

Some recent work based on genetic programming (GP) [21], originally developed to evolve computer programs, is a first step in this direction. For example, in [21, 22] neural networks have been represented as parse trees which are recombined using a crossover operator which swaps subtrees representing subnetworks. However, the graph-like structure of neural networks is not ideally represented directly with parse trees either. Indeed, an alternative approach based on GP, known as cellular encoding [23], has recognized this and used an indirect network representation in which parse trees represent rules which grow a complete network from a tiny neural embryo. Although very compact, this representation enforces a considerable bias on the network architectures that can be achieved.

Recently, a new form of GP, Parallel Distributed Genetic Programming (PDGP) [24], in which programs are represented as graphs instead of parse trees, has been applied to the evolution of neural networks [25]. The method allows the use of more than one activation function in the neural network but it does not include any operators specialized in handling the connection weights. Nonetheless, the results were encouraging and led us to believe that a PDGP-inspired representation, with specialized operators acting on meaningful building blocks, could be used to efficiently evolve neural networks. Indeed, in [26] we improved and specialized PDGP by introducing a dual representation, where a linear description of the network was dynamically associated to a two-dimensional grid. Several specialized genetic operators, some using the linear description, others using the grid, allowed the evolution of the topology and the weights of moderately complex neural networks very efficiently. In [27, 28], we proposed a combined crossover operator which allowed the evolution of the architecture, the weights and the activation function of feedforward neural networks concurrently.

In this paper, we extend the combined crossover operator to the design of recurrent networks. In the following sections, our representation and operators are described, and we report on results of the application of the paradigm to the relatively complex task of tracking and clearing a trail.

2 Representation

In PDGP, instead of the usual parse tree representation used in GP, a graph representation is used, where the functions and terminals are allocated in a two-dimensional grid of fixed size and shape. The grid is particularly useful to solve problems whose solutions are graphs with a natural layered structure, like neural networks.

However, this representation can make inefficient use of the available memory if the grid has the same shape for all individuals in the population. In fact, in this case there is no need to represent the grid explicitly in each individual. Also, in some cases, it is important to be able to abstract from the physical arrangement of neurons into layers and to only consider the topological properties of the net. To do this, it is more natural to use a linear genotype.

These limitations led us to propose [26] a dual representation, in which a linear chromosome is converted, when needed, into the grid-based PDGP representation. The dual representation includes every detail necessary to build and use the network: connectivity, weights and the activation function of each neuron. A chromosome is an ordered list of nodes (see Figure 1). An index indicates the position of the nodes in the chromosome. All chromosomes have the same number of nodes.

Like in standard genetic programming, the nodes are of two kinds: functions and terminals. The functions represent the neurons of the neural network, and the terminals are the variables containing the input to the network (see Figure 2a).

When the node is a neuron, it includes the activation function and the bias, as well as the indexes of other nodes sending signals to the neuron and the weights necessary for the computation of its output. Multiple connections from the same node are allowed.

When necessary (see Section 3), the linear representation just described is transformed into the two-dimensional representation used in PDGP. A description table defines the number of layers and the number of nodes per layer of the two-dimensional representation (see Figure 2b and c). The nodes of the linear chromosome are mapped onto this representation according to the description table. The connections of the network are indicated by links between nodes in the two-dimensional representation. The description table is a characteristic of the population and it is not included in the genotype. The two-dimensional representation may have any number of layers, each of different size. This feature may be used to constrain the geometry of the network (e.g. to obtain encoder/decoder nets) and can also be used to guide specialized crossover operators [26, 27].

The nodes in the first layer are terminals, whereas the nodes in the last layer represent the output neurons of the network. The number of input and output nodes depends on the problem to be solved. The remaining nodes, called *internal nodes*, constitute the internal layer(s), and they may be either neurons or terminals.

Although the size of the chromosome is fixed for the entire population, the neural networks represented may have different sizes. This happens because terminals may be present as internal nodes from the beginning, or may be introduced by crossover and mutation (this is discussed in Section 3). They are removed from the network during the decoding phase, which is performed before each individual is evaluated. Connections from the removed terminals are replaced with connections from corresponding terminals in the input layer (see Figures 2c and d).

Our model allows the use of more than one activation function, so that a suitable combination of activation functions can be evolved to solve a particular problem (this is discussed in the next section).

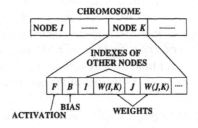

Fig. 1. Chromosome and node description.

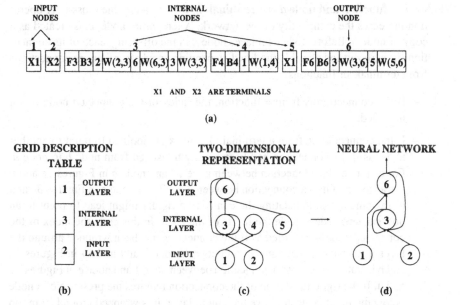

X1 AND X2 ARE TERMINALS

(a)

(b)

(c)

(d)

Fig. 2. (a) Example of a chromosome. (b) The table describing the number of nodes per layer of the network. (c) The two-dimensional representation resulting from the mapping of the nodes of the chromosome in (a), according to the description table in (b) (note the intron in the internal layer). (d) Resulting neural network after elimination of introns and transfer of connection from terminals in the internal layer to corresponding terminals in the first layer (Note that node 5 is the same terminal as node 1).

3 Crossover

By experimenting with different combinations of crossover and mutation operators in our previous work [26], we have drawn the conclusion that, for the evolution of neural networks, it is important to use operators which induce a fitness landscape as smooth as possible, and that it is also important to treat connections from terminals differently from connections from functions (neurons).

The crossover operator proposed in this paper works by randomly selecting a node a in the first parent and a node b in the second parent (see Figure 3a), and by replacing node a in a copy of the first parent (the offspring). Depending on the types of node a and node b, the replacement of node a is carried out as follows:

Both nodes are terminals This is the simplest case, node b replaces node a, and there is no change either in the topology or in the weights of the network.

Node b is a terminal and node a is a function In this case, node b also replaces node a, but the complexity of the network is reduced, because a neuron is removed from the network.

Node b is a function and node a is a terminal In this situation, the cross-over operation increases the complexity of the network. A temporary node, c, is created as a copy of node b. Before node c replaces node a in the offspring, each of its connections is analyzed and possibly modified, depending on whether they are connections from terminals or functions.

- If the connection is from a function, the index of the connected node is not modified.
- If the connection is from a terminal, the index is modified to point to another node, as if the connection had been rigidly translated from node b to node a. For example, the connection between node 10 and node b in Figures 3a and b is transformed into a connection between node 7 and node c in Figures 3c and d. In some cases, translating the connection rigidly might lead to point to an non-existent node outside the limits of the layer. In this case, the index of the connected node is modified as if the connection had been wrapped around the layer. For instance, the connection between node 8 and node b in Figures 3a and b is transformed into a connection between node 1 and node c in Figures 3c and d. If the rigid translation of the connection requires the presence of a node below the input layer or above the output layer, it is wrapped around from top to bottom (or vice-versa), as if the columns were circular.

This procedure for connection inheritance aims at preserving as much as possible the information present in the connections and weights.

Both nodes are functions This is the most important case. By combining the description of two functions, the topology and the weights of the network can be changed. After creating node c as described above, its description and the description of node a are combined by selecting two random crossover points, one in each node, and by replacing the connections to the right of the crossover point in node a with those to the right of the crossover point in node c, thus creating a new node to replace node a in the offspring. See Figure 4.

This process can easily create multiple connections between the same two nodes. These are very important because their net effect is a fine tuning of the connection strength between two nodes. However, as this may reduce the efficiency of the search, we only allow a prefixed maximum number of multiple connections. If more than the allowed maximum number of multiple connections are created, some of them are deleted before the replacement of node a in the offspring.

Modification of the activation function and bias of a node is not performed with our crossover operator. However, this can be indirectly accomplished by replacing a function with a terminal, which can later be replaced with a function with different features.

This crossover operator can not only evolve the topology and strength of the connections in a network, but also the number of neurons and the neurons themselves, by replacing their activation functions and biases. Similarly to GP, even if both parents are equal, the offspring may be different. This helps to keep diversity.

135

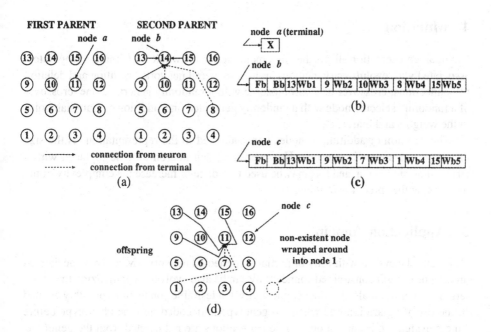

Fig. 3. (a) Two-dimensional representation of the parents. For clarity, only connections relevant to the operation are shown. (b) Nodes *a* and *b*. (c) Node *c* is a copy of node *b* with modified connections. The connections of node *b* whose indexes indicated connections from terminals received new indexes. (d) Offspring generated by replacing node *a* with node *c*.

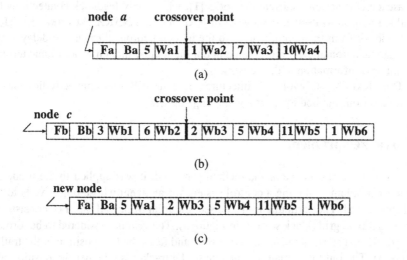

Fig. 4. Combination of two functions. (a) Node *a*. (b) Node *c* . (c) New node created to replace node *a* in the offspring. Note the multiple connections created.

4 Mutation

Our dual representation allows the implementation of the whole set of mutation operators associated with evolutionary methods applied to neural networks: addition and deletion of connections, crossover of a chromosome with a randomly generated one, crossover of a randomly selected node with a randomly generated one, addition of Gaussian noise to the weights and biases, etc..

The deletion or addition of nodes is not allowed in the representation, as the size of the chromosomes is constant. However, a hidden neuron may be replaced with a terminal or vice-versa, and this may be used to reduce or increase the complexity of the network, within predefined limits.

5 Application domain

Our method can deal with fully or partially connected networks. Moreover, specialized architectures with constrained connectivity can be also evolved by initializing the population with individuals of proper connectivity. Undesirable connections possibly created in the offspring are ignored when the genotype is decoded into the phenotype before fitness evaluation is carried out. These connections are not deleted from the genotype, they are introns not expressed in the phenotype, and may become active again later.

For example, in Elman networks [29] the context units can be represented by feedback connections from the hidden neurons to themselves. The remaining connections in the network are feedforward connections. If connections not complying with these constraints are created, they can be ignored during network evaluation. The same applies to cascade correlation recurrent networks [1], where only feedback connections from a hidden neuron to itself are allowed. In NARMA and NARX networks [30, 31, 32] the feedback from the output neuron to the hidden neurons through the delayed input units can be automatically represented in our method by including additional terminals to input this information to the network.

This flexibility to evolve architectures of quite different connectivities makes it possible for our method to cover a wide range of tasks.

6 Tracker problem

To assess the performance of the method proposed, it was applied to the moderately complex problem of finding a control system for an agent whose objective is to track and clear a trail. The particular trail we used, the John Muir trail [33, 11], consists of 89 tiles in a 32x32 grid (black squares in Figure 5). The grid is considered to be toroidal.

The tracker starts in the upper left corner, and faces the first position of the trail (see Figure 5). The only information available to the tracker is whether the position ahead belongs to the trail or not. Based on this information, at each time step, the tracker can take 4 possible actions: *wait* (doing nothing), *move forward* (one position), *turn right 90°* (without moving) or *turn left 90°* (without moving). When the tracker moves to a position of the trail, that position is immediately cleared. This is a variant of the well known "ant" problem often studied in the GP literature [21].

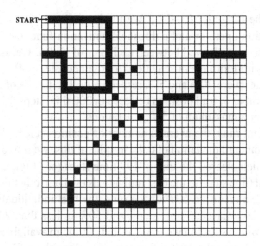

Fig. 5. John Muir trail.

Usually, the information to the tracker is given as a pair of input data [33, 11]: the pair is (1,0) if the position ahead of the current tracker position belongs to the trail, and (0,1) if it does not. The objective is to build a neural network that, at each time step, receives this information, returns the action to be carried out, and clears the maximum number of positions in a specified number of time steps (200 in our experiments). As information about where the tracker is on the trail is not available, it is clear that to solve the problem the neural network must have some sort of memory in order to remember its position. As a consequence, a recurrent network is necessary. Although it might seem to be unnecessary, the *wait* action allows the network to update its internal state while staying at the same position (this can be imagined as "thinking" about what to do next).

This problem is very hard to solve in 200 time steps (in [21], John Koza allotted 400 time steps to follow slightly different trails using GP). However, it is relatively easy to find solutions able to clear up to 90% of the trail (in [33] Jefferson et al. present a statistical evaluation of this subject using finite automata and neural networks). This means that the search space has many local minima which mislead evolution, as confirmed by Langdon & Poli [34], who studied the fitness landscape of the ant problem for GP. This makes the problem a good benchmark to test the ability of our method to evolve recurrent neural networks.

7 Experimental Results

We used asymmetric recurrent neural networks to solve the problem. A neuron can receive connections from any other neuron (including output neurons). All neurons are evaluated synchronously as a function of the output of the neurons in the previous time step, and of the current input to the network. Initially, all neurons have null output.

In all experiments a population of 100 individuals was evolved. All individuals were initialized with 10 internal nodes in a 2-10-2 grid. The weights and biases were randomly initialized within the range [-1.0, +1.0]. We used crossover with a randomly created individual as mutation operator and a threshold activation function. A maximum of 5 multiple connections was allowed between each pair of nodes. We used a generational genetic algorithm with tournament selection (tournament size = 4). The crossover and mutation probabilities were 70% and 5%, respectively. The fitness of an individual was measured by the number of trail positions cleared in 200 time steps. The population was allowed to evolve for a maximum of 500 generations.

In 20 independent runs, the average number of positions cleared by the best individuals was 81.5 (standard deviation of 6.4). In spite of this, we found a network with 7 hidden neurons, which cleared the trail in 199 time steps (see Figure 6). Moreover, by assigning additional time steps to the best individuals evolved, other networks were able to clear the trail. As a result, in 60% of the runs, the best individuals evolved cleared the 89 positions in less than 286 time steps, and 50% in less than 248 time steps. An unexpected feature of the solution we found is that, although available, it does not make use of either the *turn left* or the *wait* options to move along the trail.

These are very promising results. For comparison, Jefferson et al. [35] report a solution with 5 hidden neurons, which clears the trail in 200 time steps. The solution is a hand-crafted architecture trained by a genetic algorithm using a huge population (65,536 individuals). Using an evolutionary programming approach, Angeline et al. [11] report a network with 9 hidden neurons, evolved in 2090 generations (population of 100 individuals). The network clears 81 positions in 200 time steps and takes additional 119 time steps to clear the entire trail. They also report another network evolved in 1595 generations, which scores 82 positions in 200 time steps.

In order to show the ability of our method to evolve neural networks with more than one activation function, 20 additional independent runs were carried out with two activation functions: the threshold activation function of the previous experiment and the hyperbolic tangent. We found a network with 6 hidden neurons and 56 connections able to clear the trail in 259 steps.

8 Conclusion

In this paper, a new approach to the automatic design of recurrent neural networks has been presented, which makes natural use of their graph structure. The approach is based on a dual representation and a new crossover operator. The method was applied to evolve asymmetric recurrent neural networks with promising results.

In future research, we intend to extend the power of the method, and to apply it to a wider range of practical problems.

9 Acknowledgements

The authors wish to thank the members of the EEBIC (Evolutionary and Emergent Behavior Intelligence and Computation) group for useful discussions and comments. This research is partially supported by a grant under the British Council-MURST/CRUI agreement and by CNPq (Brazil).

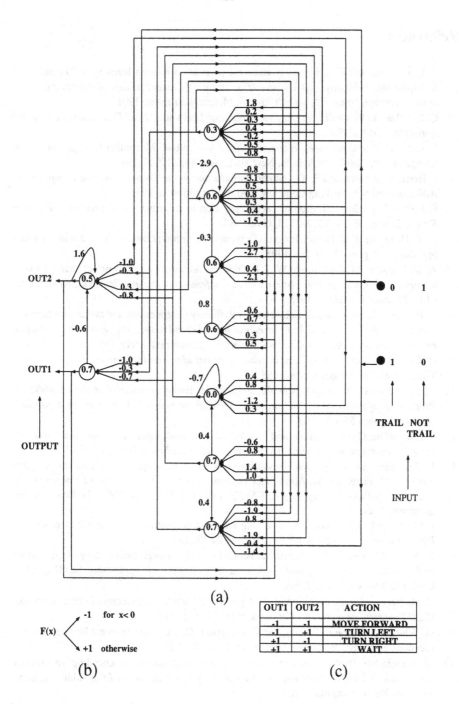

(a)

$$F(x) \begin{cases} -1 & \text{for } x < 0 \\ +1 & \text{otherwise} \end{cases}$$

(b)

OUT1	OUT2	ACTION
-1	-1	MOVE FORWARD
-1	+1	TURN LEFT
+1	-1	TURN RIGHT
+1	+1	WAIT

(c)

Fig. 6. Full solution to the tracker problem. (a) Neural network. Values in the circles are the biases. (b) Activation function used. (c) Mapping of the network output into actions.

References

1. S. E. Fahlman and C. Lebiere. A recurrent cascade-correlation learning architecture. In R. Lippmann, J. Moody, and D. Touretzky, editors, *Advances in Neural Information Processing Systems*, volume 3, pages 190–196. Morgan Kaufmann, 1991.

2. C. L. Giles and W. Omlin. Pruning recurrent neural networks. *IEEE Transactions on Neural Networks*, 5(5):848–855, 1994.

3. S. Haykin. *Neural networks, a comprehensive foundation.* Macmillan College Publishing Company, Inc., 866 Third Avenue, New York, New York 10022, 1994.

4. J. Hertz, K. Anders, and R. G. Palmer. *Introduction to the Theory of Neural Computation.* Addison-Wesley Publishing Company, Redwood, California, 1991.

5. F. J. PINEDA. Generalization of backpropagation to recurrent neural networks. *Physical Review Letters*, 59(19):2229–2232, Nov. 1987.

6. D. R. Hush and B. G. Horne. Progress in supervised neural networks. *IEEE Signal Processing Magazine*, pages 8–39, Jan. 1993.

7. A. M. Logar, E. M. Corwin, and W. J. B. Oldham. A comparison of recurrent neural network learning algorithms. In *IEEE International Conference on Neural Networks*, pages 1129–1134, Stanford University, 1993.

8. T. Bäck, G. Rudolph, and H. Schwefel. Evolutionary programming and evolution strategies: Similarities and differences. In *Proceedings of the Second Annual Conference on Evolutionary Programming*, pages 11–22. Evolutionary Programming Society, 1993.

9. D. Goldberg. *Genetic algorithm in search, optimization and machine learning.* Addison-Wesley, Reading, Massachusets, 1989.

10. K. Lindgren, A. Nilsson, M. Nordahl, and I. Rade. Evolving recurrent neural networks. In *Proceedings of the International Conference on Artificial Neural Nets and Genetic Algorithms (ANNGA)*, pages 55–62, 1993.

11. P. J. Angeline, G. M. Saunders, and J. B. Pollack. An evolutionary algorithm that constructs recurrent neural networks. *IEEE Transactions on Neural Networks*, 5(1), 1994.

12. J. McDonnell and D. Waagen. Neural network structure design by evolutionary programming. In D. Fogel and W. Atmar, editors, *Proceedings of the Sec. Annual Conference on Evolutionary Programming*, pages 79–89, La Jolla, CA, USA, Feb. 1993. Evolutionary Programming Society.

13. J. McDonnell and D. Waagen. Evolving recurrent perceptrons for time-series modelling. *IEEE Transactions on Neural Networks*, 5(1):24–38, Jan. 1994.

14. X. Yao and Y. Liu. Evolving artificial neural networks through evolutionary programming. In *Proceedings of the 5th Annual Conference on Evolutionary Programming*, San Diego, CA, USA, Feb/Mar 1996. MIT Press.

15. V. Maniezzo. Genetic evolution of the topology and weight distribution of neural networks. *IEEE Transactions on Neural Networks*, 5(1):39–53, 1994.

16. D. Whitley, S. Dominic, R. Das, and C. Anderson. Genetic reinforcement learning for neurocontrol problems. *Machine Learning*, 13:259–284, 1993.

17. M. Mandischer. Evolving recurrent neural networks with non-binary encoding. In *Proceedings of the 2nd IEEE Conference on Evolutionary Computation (ICEC)*, volume 2, pages 584–589, Perth, Australia, Nov. 1995.

18. J. Santos and R. Duro. Evolutionary generation and training of recurrent artificial neural networks. In *Proceedings of the first IEEE Conference on Evolutionary Computation (ICEC)*, volume 2, pages 759–763, Orlando, FL, USA, Jun. 1994.

19. S. Bornholdt and D. Graudenz. General asymmetric neural networks and structure design by genetic algorithms. *Neural Networks*, 5:327–334, 1992.

20. F. Marin and F. Sandoval. Genetic synthesis of discrete-time recurrent neural network. In *Proceedings of International Workshop on Artificial Neural Network (IWANN)*, pages 179–184. Springer-Verlag, 1993.

21. J. R. Koza. *Genetic Programming, on the Programming of Computers by Means of Natural Selection*. The MIT Press, Cambridge, Massachusets, 1992.

22. B. Zhang and H. Muehlenbein. Genetic programming of minimal neural nets using Occam's razor. In S. Forrest, editor, *Proceedings of the 5th international conference on genetic algorithms (ICGA'93)*, pages 342–349. Morgan Kaufmann, 1993.

23. F. Gruau. *Neural network synthesis using cellular encoding and the genetic algorithm*. PhD thesis, Laboratoire de L'informatique du Parallélisme, Ecole Normale Supériere de Lyon, Lyon, France, 1994.

24. R. Poli. Some steps towards a form of parallel distributed genetic programming. In *Proceedings of the First On-line Workshop on Soft Computing*, pages 290–295, Aug. 1996.

25. R. Poli. Discovery of symbolic, neuron-symbolic and neural networks with parallel distributed genetic programming. In *3rd International Conference on Artificial Neural Networks and Genetic Algorithms (ICANNGA)*, 1997.

26. J. C. F. Pujol and R. Poli. Evolution of the topology and the weights of neural networks using genetic programming with a dual representation. Technical report CSRP-97-07, The University of Birmingham, School of Computer Science, 1997.

27. J. C. F. Pujol and R. Poli. A new combined crossover operator to evolve the topology and the weights of neural networks using a dual representation. Technical report CSRP-97-12, The University of Birmingham, School of Computer Science, 1997.

28. J. C. F. Pujol and R. Poli. Evolving neural controllers using a dual network representation. Technical report CSRP-97-25, The University of Birmingham, School of Computer Science, 1997.

29. J. L. Elman. Finding structure in tima. *Cognitive Science*, 14:179–211, 1990.

30. J. T. Connors and R. D. Martin. Recurrent neural networks and robust series prediction. *IEEE Transactions on Neural Networks*, 5(2):240–254, Mar. 1994.

31. T. Lin, B. G. Horne, P. Tino, and C. L. Giles. Learning long-term dependencies in narx recurrent neural networks. *IEEE Transactions on Neural Networks*, 7(6), Nov. 1996.

32. H. T. Siegelmann, B. G. Horne, and C. L. Giles. Computational capabilities of recurrent narx neural networks. *IEEE Transactions on Systems, Man and Cybernetics*, 27(8), Mar. 1997.

33. R. Collins and D. Jefferson. Antfarm: toward simulated evolution. In C. Langton, C. Taylor, J. Farmer, and S. Rasmussen, editors, *Artificial Life II, Santa Fe Institute Studies in the Sciences of Complexity*, volume X. Addison-Wesley, 1991.

34. W. Langdon and R. Poli. Why ants are hard. Technical report CSRP-98-04, The University of Birmingham, School of Computer Science, 1998.

35. D. Jefferson, R. Collins, C. Cooper, M. Dyer, M. Flowers, R. Korf, C. Taylor, and A. Wang. Evolution as a theme in artificial life: The genesys/tracker system. In C. Langton, C. Taylor, J. Farmer, and S. Rasmussen, editors, *Artificial Life II, Santa Fe Institute Studies in the Sciences of Complexity*, volume X. Addison-Wesley, 1991.

A Cellular-Programming Approach to Pattern Classification

Giovanni Adorni, Federico Bergenti, Stefano Cagnoni*

Department of Computer Engineering, University of Parma, Italy

Abstract. In this paper we discuss the capability of the cellular programming approach to produce non-uniform cellular automata performing two-dimensional pattern classification. More precisely, after an introduction to the evolutionary cellular automata model, we describe a general approach suitable for designing cellular classifiers. The approach is based on a set of non-uniform cellular automata performing specific classification tasks, which have been designed by means of a cellular evolutionary algorithm.

The proposed approach is discussed together with some preliminary results obtained on a benchmark data set consisting of car-plate digits.

1   Introduction

Cellular Automata (CAs) have been introduced by von Neumann [1] in the early fifties as systems capable of self-reproduction and self-organization. As an informal definition, a CA is a system composed by a grid of cells each of which performs simple computations. Each cell is connected to a set of nearest neighbours in order to make information exchange through the grid possible. From the classical viewpoint, all cells compute the same function and the complex behaviour of the system emerges from the synchronous application of all identical cell rules to different data. This homogeneous behaviour makes CAs suitable for the simulation of the dynamics of complex physical systems in time [2]. In more recent years, CAs have been used in many other applications where fast, and possibly parallel, computation is required, such as in low-level vision, e.g. for inverse perspective mapping or optical flow calculation [3]. From a theoretical point of view, CAs are an interesting model for massively parallel computation [4].

One of the main problems arising in the application of CAs to real-world problems is the design of appropriate rules. In fact, it is quite difficult to map a problem requiring global computation onto a system which relies on local interactions. Sipper [5] has shown that global problems can be faced using an extension of the CA model, called non-uniform CA, in conjunction with a particular co-evolutionary algorithm. This new paradigm seems to offer advantages with respect to the conventional CA model in terms of efficiency and flexibility in the design of complex applications.

---

* Address for correspondence: Stefano Cagnoni, Dept. of Computer Engineering, University of Parma, viale delle Scienze, 43100 Parma, Italy

In this paper, we investigate the performances of co-evolutionary non-uniform CAs in pattern classification. In particular, we have applied a non-uniform CA-based classifier to a set of images of digits taken from real car plates. The reason for this choice is that this application seems to be a good benchmark as: i) it intrinsically requires a training phase as the problem is data-dependent, and ii) the performance of the CAs on this task can easily be assessed using some kind of index related to the recognition rate.

Next section introduces the evolutionary non-uniform CA model. Section 3 describes the proposed approach applied to pattern classification. The benchmark on car-plate digits is illustrated in section 4, and the results obtained are discussed in section 5. Finally, section 6 reports some comments on the results and suggests possible future developments.

2 Evolutionary CAs

The idea of using an evolutionary algorithm to find a uniform-CA rule capable of performing some computational tasks has given good results in a number of applications [6], but co-evolved non-uniform CAs can yield better performances when required to exhibit a global behaviour [5].

A non-uniform CA, or hybrid CA, can be informally defined as a CA in which a different local computation is possible for every cell in the grid. During the last few years, non-uniform CAs have been applied to many different problems including fast hardware implementation [7] and associative memories [8]. Non-uniform CAs add a new degree of freedom to CA applications, while preserving the most interesting properties of uniform CAs, such as synchronization, parallelism and locality of connections between computing elements. Non-uniform CA implementation is straightforward, as it can be stated that for every non-uniform CA there exists a uniform one on which it can be mapped.

Each cell of a non-uniform CA can compute a different function; this allows for a better exploitation of the local properties of the problem. However, with respect to the uniform paradigm, non-uniform CAs require that a different rule be designed for each cell in the grid, thus enlarging the rule space. This is the reason why automatic design procedures such as evolutionary learning are the method of choice for non-uniform CA design. In this work, we adopted the "cellular genetic algorithm" (CGA) described in [9], in which the rule for each cell is evolved as a single-individual population, allowing for migration and recombination between neighbouring cells. This results in the co-evolution of a set of rules driven by some sort of local fitness.

The CGA has been used to solve, among others, two theoretical problems known as the density task and the synchronization task [5]. In the density task all cells of a one-dimensional two-state CA are required to be driven to the state which is predominant in a given initial state of the automaton. In the synchronization task the problem is to synchronize all cells to jointly flip their (identical) state at each step, starting from a random initial state. The previous problems are two global tasks: the solutions are expressed in terms of the global properties of the entire grid. Moreover, they do not allow for an easy solution to be found with the uniform-CA paradigm. The fact that they could be handled quite easily with the CGA suggests that such a method may also be suitable for more complicated tasks.

A possible limitation of CGAs, that prevents them from being a general-purpose tool, is that fitness is considered a local property of each cell, and no global fitness is defined for the global result of the computation. However, such a limitation allows the evolution to be implemented directly on cellular hardware. This property permits the whole system to process data keeping the evolution running, thus satisfying the most basic requirement for evolvable hardware [9].

3 Co-evolved CAs and Pattern Classification

The result of a CA computation is a set of values, corresponding to the states in which each cell is found after a settled number of steps.

In the application discussed in this work, CGA is used as an evolutionary supervised learning algorithm for two-dimensional pattern classification: a two-state non-uniform CA is fed with a training set of pre-classified patterns along with the corresponding labels. Therefore, in order to use a two-state CA as a binary classifier, a possibility is to choose one particular cell and use its final state as the classifier output. Given a class of patterns to be recognized, a cell is required to be "on" in the presence of a pattern belonging to that "target" class and "off" otherwise. The local fitness is defined as the root mean square of the cell sensitivity and specificity:

$$fitness = \sqrt{\frac{(Sensitivity^2 + Specificity^2)}{2}}$$

where *Sensitivity* is computed on the whole training set as the frequency with which the cell has reached the "on" state in the presence of patterns belonging to the target class; *Specificity* is estimated as the frequency with which the cell has reached the "off" state in the absence of the target class. The use of the mean square root ensures that a good balance between *Specificity* and *Sensitivity* is reached.

This fitness is local, as it is measured only on the final state of each cell; therefore the CGA can be used to evolve the CA in order to maximise the fitness of all cells. Such a training phase generally results in a CA organised in clusters of cells characterised by the same rule (quasi-uniform CA) and giving similar performances.

To exploit the CA for classification, the final state of the cell with the highest fitness, or a suitably chosen set of well-performing cells can be used.

An ideal classification could be achieved by a set of CAs, each trained to recognize a specific class, with *Sensitivity* = *Specificity* = 1. In this situation, each input pattern would activate one and only one classifier. As this is never the real case, the output of all classifiers must be taken into account to classify an input pattern. Therefore, the operation of the CA-based classifiers can be regarded as a transformation from the input pattern space to a reduced-size output space. To reach the final classification, a decision criterion has to be defined in the output space (see section 4.2).

4 Classification of Digits in Car Plates

In order to test the co-evolved non uniform CA paradigm on a non-trivial classification task, a database of digits derived from the acquisition of real car-plate images from different viewpoints and in different lighting conditions has been considered. Each digit consists of an 8 × 13 pixel gray-scale pattern.

A simple preprocessing stage has been applied to each digit to obtain binary patterns that can be handled by two-state CAs. Such a stage consists of histogram stretching followed by thresholding. Some examples of the preprocessed patterns are shown in fig. 1.

Fig. 1. Examples of the original (top) and preprocessed (bottom) patterns

The classification of the binary pattern is obtained through a CA-based architecture designed with the aim of making parallel implementation on a single cellular machine possible.

4.1 CA-based architecture

In our implementation, each CA consists of a grid of the same size as the input pattern (8 × 13) and it is run for 24 steps, that is long enough to allow information to propagate through the grid. The von Neumann neighbourhood is adopted and the grid is periodic in space, that is the leftmost cell in each row is connected to the rightmost one, and the down-most cell in each column is connected to the up-most one.

As we have anticipated, to obtain a CA-based classification of the patterns considered, the most straightforward architecture consists of 10 CAs separately trained to detect the different digits. Such an architecture was the starting point in the development of our pattern classifiers.

During the training phase, each CA-based classifier was evolved through 5 runs of the CGA, iterated for 20,000 generations with a mutation rate of 0.001 on

Digit	Sensitivity	Specificity	Fitness
0	0.964	0.995	0.980
1	0.873	0.980	0.928
2	0.932	0.945	0.938
3	0.894	0.968	0.932
4	0.937	0.950	0.943
5	0.925	0.963	0.944
6	0.958	0.925	0.942
7	0.989	0.977	0.984
8	0.840	0.976	0.911
9	0.880	0.967	0.925

Table 1. Performances of the digit classifiers

a training set of 892 patterns. The best CA for each digit is used for classification using the final state of the classifier's highest-fitness cell as a binary output.

Table 1 shows the results obtained for each classifier on a test set of 884 patterns.

The dimensionality reduction of the input space produced by this architecture is so relevant that it tends to produce a very high number of classification collisions, deriving from the mapping of input patterns which belong to more than one class onto the same output pattern.

As a preliminary test aimed at evaluating the discrimination capabilities of this set of classifiers we estimated the maximum achievable performance compatible with the presence of collisions. As the estimated performance was quite low (about 90%) we decided to extend this simple architecture to improve it.

In this regard, the information that can be obtained from the analysis of the results of the classifier set previously described can be used to design a more specialized architecture, capable of reducing the number of collisions. To do so, a number of classifiers trained to solve the most critical cases can be added. Furthermore, it can be expected that the new classifiers achieve rather good performances, as a binary classifier generally works better if the features of the two classes it has to discriminate are as uniform as possible. This happens, for example, if a classifier has to distinguish only between two classes, as demonstrated by the results shown in table 2.

Based on these assumptions, we extended the architecture by adding a set of classifiers trained on patterns belonging to the pairs of classes that produced the highest number of collisions. This resulted in a new 19-classifier architecture.

4.2 Decision Criterion

The problem of deriving a final decision from a set of uncertain ones, such as the output of the 19 CA-based classifiers, also known as multi-expert classification, has been tackled with several different approaches. One of the most easily

Digit Pair	Sensitivity	Specificity	Fitness
0/9	1.000	0.988	0.994
7/1	0.966	0.988	0.977
2/3	0.977	1.000	0.989
5/6	0.966	0.972	0.969
9/6	0.976	0.986	0.981
6/8	0.958	1.000	0.979
4/1	0.989	0.977	0.983
1/3	1.000	0.990	0.995
3/8	0.981	1.000	0.990

Table 2. Results of the pairwise classifiers

implementable methods, as it does not require much additional training, is the so-called stacked generalizer [10].

The idea of this method is to feed all classifiers with the training patterns: for each configuration of the classifiers' output the histogram of the frequencies with which the configuration has been associated to each class is computed. After this simple phase, the information contained in such histograms can be regarded as an estimation of the conditioned probability of obtaining a given output when a pattern of a certain class is presented as input.

Classification has a very high degree of certainty when a configuration of the classifiers' output has occurred a sufficient number of times during the training phase and the histogram shows that a class is predominant over all others. In this case the label associated with the predominant class can be assigned to the input pattern which produces such a configuration. When the output configuration has occurred too few times to be statistically significant or when the histogram associated with it tends to be multi-modal, a proximity criterion with respect to a set of class prototypes can be used. This strategy to select how to perform the final classification is based on the values of the two highest peaks of the histogram and can be summarized as follows:

```
if (highest peak < A) or ((highest peak / second highest) < B)
then
    class = closest prototype label
else
    class = highest peak label
end if
```

where A and B are two thresholds whose values are to be preset or experimentally determined. In our case, the proximity criterion is based on a vector codebook obtained using the OSLVQ algorithm (Optimum Size Learning Vector Quantization) [11], a variation of the well-known LVQ [12] aimed at optimising the codebook size.

5 Results

The last step in the design of the architecture is the definition of the stacked generalizer parameters (the two thresholds A and B). In our experiments, A was set to 0.5% of the training set size and B was set to 50.

To assess the performances of the CA-based architecture, we used it to classify the 884 patterns comprised in the test set, obtaining a percentage of correct classifications of about 96.7%.

The confusion matrix for the CA-based classifier is reported in table 3. In the matrix, the elements C_{ij} with $i = j$, indicate the percentage of patterns belonging to class i that have been correctly classified; elements C_{ij} with $i \neq j$, indicate the percentage of patterns belonging to class j that have been misclassified as belonging to class i.

	0	1	2	3	4	5	6	7	8	9
0	**0.964**	0.000	0.011	0.000	0.000	0.000	0.000	0.000	0.000	0.012
1	0.000	**0.978**	0.000	0.000	0.000	0.000	0.000	0.223	0.000	0.000
2	0.000	0.000	**0.932**	0.000	0.000	0.000	0.000	0.000	0.000	0.000
3	0.000	0.000	0.000	**0.990**	0.000	0.037	0.000	0.000	0.000	0.000
4	0.000	0.011	0.023	0.000	**1.000**	0.000	0.000	0.000	0.000	0.000
5	0.012	0.000	0.000	0.010	0.000	**0.925**	0.000	0.000	0.000	0.000
6	0.012	0.000	0.000	0.000	0.000	0.029	**0.972**	0.000	0.000	0.000
7	0.000	0.011	0.034	0.000	0.000	0.000	0.000	**0.977**	0.000	0.024
8	0.012	0.000	0.000	0.000	0.000	0.029	0.028	0.000	**0.987**	0.024
9	0.000	0.000	0.000	0.000	0.000	0.000	0.000	0.000	0.013	**0.940**

Table 3. Confusion matrix for the CA-based classifier

More than 60% of the patterns could be classified directly using the stacked generalizer. The remaining 40% were classified by means of the codebook vector obtained using the OSLVQ algorithm. As the number of the training patterns used is relatively small with respect to the complexity of the problem, it can be expected that increasing the size of the training set will produce some improvements: the number of patterns that can be directly classified without resorting to the proximity criterion is likely to grow, along with the global performances of the architecture.

By comparing the results of the final architecture to the ones obtained by the single-digit classifiers, the importance of the specialized classifiers becomes clear. In fact, as one could expect, the worst performances of the single-digit classifiers were obtained by the ones associated to morphologically similar patterns, e.g. 3, 8, and 9. After adding the pairwise classifiers a great improvement in the classification of such critical patterns was obtained. This is particularly evident in the case of the classification of the digit 8 for which the value of the corresponding sensitivity (diagonal) element is 0.987, compared to a sensitivity of 0.84 for the single-digit classifier.

Fig. 2. Sample output of the recognition system based on the cellular classifier.

An example of application of the system, extended also to letters, for car-plate recognition, is shown in figure 2.

6 Conclusions

Obtaining an emerging global behaviour from a locally connected computational model is a difficult design task. In this paper we have investigated the capabilities of co-evolved non-uniform CAs to perform pattern classification, which is generally considered a non-trivial problem requiring global computation.

The described results, despite being preliminary and obtained using a relatively small training set, appear to be encouraging. In particular, the CA-based architecture is able to perform an efficient dimensionality reduction of the pattern space while giving satisfactory performances.

The architecture seems to provide a good trade-off between efficiency and performance. However, given the results obtained by adding new classifiers to solve classification conflicts, it can be expected that, if performances are the main concern, a slightly larger and less efficient architecture would yield better results. The work is moving in this direction and the first results obtained seem to confirm this hypothesis.

The procedure through which the CA-based architecture has been designed is very general and can be applied to virtually all classification tasks.

Finally, it is worth noting that although the work has been described in terms of the non-uniform CA paradigm, the CAs have been implemented on a CAM-8 computer, designed by the Information Mechanics Group of MIT [13], which relies upon a uniform CA model.

Acknowledgements

The authors thank Marco Gori for his helpful suggestions during the preparation of the work and for supplying them with the database of car-plate images.

References

1. J. von Neumann, *Theory of Self-Reproducing Automata*, University of Illinois Press, 1966.
2. T. Toffoli and N. Margolus, *Cellular Automata Machines: A New Environment for Modeling*, MIT Press, Cambridge, Massachusetts, 1987.
3. G. Adorni, A. Broggi, G. Conte, and V. D'Andrea, "A self tuning system for real-time optical flow detection", in *Proc. IEEE System, Man, and Cybernetics Conference*, 1993, vol. 3, pp. 7–12.
4. S. Wolfram, *Theory and applications of cellular automata*, World Scientific, Singapore, 1986.
5. M. Sipper, *Evolution of Parallel Cellular Machines: the Cellular Programming Approach*, Springer-Verlag, Berlin, 1997.
6. M. Mitchell, J.P. Crutchfield, and R. Das, "Evolving cellular automata with genetic algorithms: A review of recent work", in *Proceeding of the first International Conference on Evolutionary Computation and its applications (EvCA'96)*, 1996.
7. P.P. Chaudhuri, D.R. Chowdhury, S. Nandi, and S. Chattopadhyay, *Additive Cellular Automata: Theory and Applications*, IEEE Computer Society Press, Los Alamitos, CA, 1997.
8. M. Chady and R. Poli, "Evolution of cellular-automaton-based associative memories", Tech. Rep. CSRP-97-15, University of Birmingham, School of Computer Science, May 1997.
9. M. Tomassini, "Evolutionary algorithms", in *Towards Evolvable Hardware: The Evolutionary Engineering Approach*, E. Sanchez and M. Tomassini, Eds., Berlin, 1996, LNCS, pp. 19–47, Springer.
10. D.H. Wolpert, "Stacked generalization", *Neural Networks*, vol. 5, pp. 241–259, 1992.
11. S. Cagnoni and G. Valli, "OSLVQ: a training strategy for optimum-size Learning Vector Quantization classifiers", in *Proc. IEEE International Conference on Neural Networks*, June 1994, pp. 762–765.
12. T. Kohonen, *Self-organization and associative memory (2nd ed.)*, Springer-Verlag, Berlin, 1988.
13. N. Margolus, "CAM-8: a computer architecture based on cellular automata", in *Pattern Formation and Lattice-Gas Automata*, A. Lawniczak and R. Kapral, Eds. American Mathematical Society, 1994.

Evolving Coupled Map Lattices for Computation

Claes Andersson[1] and Mats G. Nordahl[1,2]

[1] Institute of Theoretical Physics, Chalmers University of Technology, S-412 96 Göteborg, Sweden, e-mail: claesand@fy.chalmers.se

[2] Santa Fe Institute, 1399 Hyde Park Road, Santa Fe, NM 87501, USA. e-mail: mgn@santafe.edu

Abstract. Genetic programming is used to evolve coupled map lattices for density classification. The most successful evolved rules depending only on nearest neighbors ($r = 1$) show better performance than existing $r = 3$ cellular automaton rules on this task.

Distributed information processing in a spatially extended network of processing units is a common phenomenon in nature, with the human brain as the foremost example. Evolutionary algorithms have been used successfully to design algorithms for distributed computation with discrete processing units, in particular for cellular automata (CA) [19], both with homogenous (e.g. [12]) and inhomogeneous [18] dynamics. Here we apply similar methods to networks with continuous-valued processing units, which are likely to be more relevant to natural systems. In particular we consider homogenous systems on a regular lattice with local interactions, so called *coupled map lattices* (e.g., [7]), which are dynamical systems with discrete time and space but continuous states.

Computation in cellular automata has been studied from different perspectives — CA have been shown capable of universal computation (e.g. [11]), and CA have also been constructed for various specific computational tasks (e.g., [15]). Coupled map lattices have been studied as dynamical systems [7], but computation in coupled map lattices has not been studied extensively (see [16, 6] for exceptions), and is in itself an interesting topic for further study. In this contribution, we use genetic programming to evolve coupled map solutions to the task of density classification, which has been extensively studied in the cellular automaton case. Some of the solutions found use the continuous degrees of freedom in interesting ways, and also perform better than known CA rules on this task.

1 Density Classification

We have studied the problem of determining which of two symbols is most abundant in a string, or the *density classification* problem with threshold $\alpha = 1/2$ (more generally one could determine if the density of some symbol is greater than a threshold α). This is a simple example of a problem which requires global information exchange, and which is non-trivial to solve using a dynamical system with only local interactions, such as a cellular automaton or a coupled map lattice. It can also be viewed as recognition of a simple non-regular language.

Several authors have evolved cellular automaton rules for solving this problem, beginning with Packard [17], whose main purpose was to investigate whether dynamical systems performing non-trivial computational tasks tend to be close to a transition between chaotic and non-chaotic behavior (the so called "edge of chaos"). Mitchell, Crutchfield and coworkers (e.g., [12–14]) have studied the density classification problem for CA extensively, and have begun to develop a theoretical framework for explaining the computational mechanisms of the evolved rules. Koza applied genetic programming [8] and considerable computational effort to the same task, and found a rule which has the best classification accuracy achieved so far — according to [1] this rule has an accuracy of 82.326% measured on a set of 10^7 strings of length 149 drawn from an uncorrelated and unbiased ensemble.

The performance of the rules discovered can also be compared to hand-designed rules. The Gacs-Kurdyumov-Levin (GKL) rule [5] given by the simple expression

$$x_i^{t+1} = \text{majority}(x_i^t, x_{i-1}^t, x_{i-3}^t) \text{ if } x_i^t = 0$$
$$= \text{majority}(x_i^t, x_{i+1}^t, x_{i+3}^t) \text{ if } x_i^t = 1$$

was originally introduced to study questions of broken ergodicity in statistical mechanics, but also shows excellent performance on the related task of density classification. Other hand-designed rules with somewhat improved performance have been constructed by Das [3] and Davis [4].

All these CA rules use two states and a neighborhood radius of three. The initial state is the string itself, and the criterion for classification is that after a certain number of iterations, the CA should reach a fixed point which is a homogenous configuration consisting of the local state which initially is in majority. When using a dynamical system for computation, input data are either encoded in the initial state or fed to the system during the computation. The dynamical system is iterated for some time, and the result of the computation is obtained by decoding the final state of the computation. The simple choice of encoding and decoding functions above ensures that no significant computation is hidden in these steps.

With this acceptance criterion, no CA rule can solve the problem exactly on all lattice sizes (e.g., [3,9]). This can be understood by noting that for a rule which could solve the task exactly, a configuration can never be mapped from a majority of one state to a majority of the other. But if the rule solves the problem exactly on one lattice size, this can be used to construct such a mapping of a configuration on a larger lattice, by combining a large region of density just below threshold with a smaller region of maximal density. This argument does not immediately give any performance bounds; this is an important topic for further study.

For other acceptance criteria, e.g., checking whether *any* finite block of a certain kind is present, or by iterating a second CA rule, even a two-state rule with neighborhood radius one can solve this task (in particular elementary CA rule 184 in Wolfram's notation [19], see, e.g., [2]). With the acceptance criterion above, no cellular automaton rule with two states and neighborhood radius one gives reasonable performance on this task.

For a coupled map lattice, the continuous degrees of freedom allow a wider variety of complex behavior even with radius one, and we have considered solutions in terms of CML rules with radii from one up to three. A radius of r obviously limits the maximal speed of information propagation to r lattice sites per time step. When comparing rules we have compensated for this by allowing proportionally longer time for the rules of smaller radius.

2 Genetic Representation and Selection Mechanism

The computation is performed by a coupled map lattice defined by a local transition function $\phi : R^3 \to R$, which defines the mapping of a local neighborhood:

$$x_i^{t+1} = \phi(x_{i-1}^t, x_i^t, x_{i+1}^t),$$

where x_i^t is the state of lattice site i at time t. To update the lattice, this function is applied simultaneously to every neighborhood of the lattice. In other words, a coupled map lattice is a cellular automaton with continuous states. Periodic boundary conditions were used in all our simulations.

The initial state is given as a sequence of values +1 and -1. The system is iterated for a predetermined number of time steps, and an initial state is considered correctly classified if at each site the CML converges to the majority element in the initial state within a small accuracy parameter $\epsilon > 0$.

We have considered several different genetic representation schemes for the transition function ϕ. In the genetic programming representation considered here, the function ϕ is represented as $\phi(x_{-1}, x_0, x_1) = \tanh(\beta f(x_{-1}, x_0, x_1))$ so that states belong to the interval $[-1, 1]$. The function $f(x_{-1}, x_0, x_1)$ is then evolved using genetic programming. In this contribution we only consider the case where f is a polynomial.

We use a library of genetic operators that handle programs in reverse Polish notation (Genetic Forth, or **GF**). In this application the programs are polynomials in $2r + 1$ variables, and thus only a limited number of operators and variables are used. Other operators could however easily be introduced. A genome is stored as an expression in reverse Polish notation, such as

C,A,A,B,C,C,B,-0.421430,0.197510,B,+,+,+,*,-0.323140,-0.142090,
-0.098570,+,*,*,-0.219470,*,0.246370,+,+,*,*,*,+,A,+

which is the genome of the best evolved rule discussed further in Sect. 3. Here A, B and C represent the three neighborhood sites for $r = 1$. The genome can of course also be represented as a tree, see Fig. 2.

There are three types of elements in an expression: operators, constants and variables. A pre-defined operator is represented by a character; in this application we only use two operators: + and *. An operator pops m elements from the stack, performs an operation and pushes n elements back. Constants and variables simply push their associated values onto the stack. Valid expressions never empty the stack, and end with a single element on the stack. Evaluation takes place by executing the elements in order. When a valid expression has been executed, the result of the evaluation is the single value present on the stack.

The genetic operators acting on the expressions are point mutations, neutral insertions, and recombination.

A point mutation changes an element to another element of the same type. For operators and variables, a replacement is simply chosen at random from a finite set of alternatives, while constants are mutated by adding a random value drawn from a rectangular distribution over an interval $[-A, A]$. Other mutations such as multiplication with a random constant were also tried. For constants, fairly high mutation rates (0.1–0.2) were used, together with a small interval A, such as $A = 0.1$; other mutation rates were significantly smaller.

Neutral insertions are additions to the genome that alter the length of the genome without affecting the phenotype (i.e., the function viewed as a mapping $R^3 \rightarrow R$). Even though the phenotype does not change immediately, neutral mutations increase the search space, and allow additional mutations to introduce new functionalities that cannot be represented in the smaller genome space. Neutral mutations have been used in evolutionary algorithms, e.g., in [10]. Viewing the expression as an evaluation tree, insertions are accomplished by replacing leaves with entire subtrees. The rate of neutral mutations was typically 0.01 per locus in the experiments. A variable X can be replaced by either 1.000000,X,*

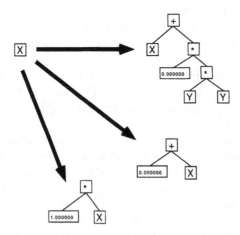

Fig. 1. The three possibilities for neutral replacements of a variable X.

or 0.000000,X,+ or X,0.000000,Y,Y,*,*,+, where Y is a randomly chosen variable. In the last case, the degree of the polynomial is increased. A constant C is replaced by C,0.000000,X,*,+, where X is a random variable.

Recombination replaces a valid subexpression (subtree) of one parent with a valid subexpression (subtree) from the other parent.

The fitness of a rule is given by its negative average error, i.e., the average over training examples of $\frac{1}{N}\sum_i(x_i^T - y_i)^2$, where N is the lattice size and y_i is the desired output (+1 or -1). For each generation, the training set is generated at random with some distribution of initial densities. In most training runs we started with a uniform distribution on a fairly wide interval of initial densities, which was gradually narrowed during the evolution.

Since the fitness of each rule is evaluated on a rather small set of examples, the noise in the fitness evaluation can be very important. This problem was addressed by using a selection algorithm similar to that of Mitchell et al [13], where a number n_E of elite rules are copied without modification to the next generation (with n_E equal to 5% or 10% of the population size), and parents are chosen randomly in the elite group. In this way fitness fluctuations can be reduced through multiple evaluations of surviving rules. The noisy fitness evaluation is still a problem for the GP algorithm; the use of elite selection together with neutral mutations also requires careful tuning of parameters.

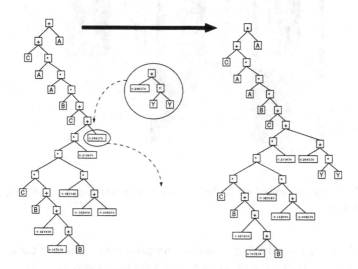

Fig. 2. The best evolved rule discussed in Sect. 3, together with the action of a neutral insertion at the locus of a constant.

3 Results

Fig. 3 shows the time evolution of the fitness in a typical GP run, in this case for radius $r = 1$. The curve shows the average squared error of the best rule in each generation, evaluated on a separate test set consisting of 1000 examples

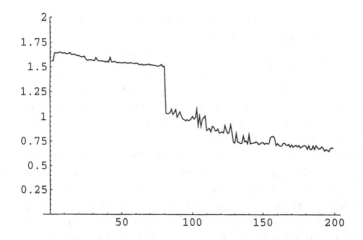

Fig. 3. Time evolution of the average classification error for a sample run.

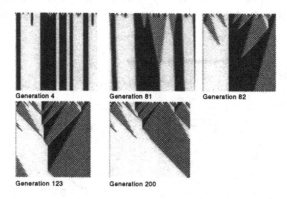

Generation 4   Generation 81   Generation 82

Generation 123   Generation 200

Fig. 4. Examples of simulations of the best rules at generations $t = 4, 81, 82, 123, 200$ in the run above.

generated with density 0.5. Fig. 4 shows examples of simulations of the best rules at certain points in the evolution (with identical initial states).

In this case the population size was 100, and in each generation a new training set with 100 examples was used. These were generated from a uniform distribution of densities in an interval $[t/200, 1 - t/200]$ that decreased linearly with time. A lattice of size 99 was used for training (the ability to generalize to larger lattice sizes was then tested for the best solutions). The constant β was set to 1.5. The initial state was created by generating two elements (excluding operators) at random, and then generating elements at random until stack level 1 had been reached twice.

The polynomials of the best rules at times $t = 0, 40, 80, 120, 160, 200$ are shown in Table 1. In the initial phase, the polynomials are linear, which for $r = 1$ gives rather poor results (for $r = 3$, linear rules can perform significantly

Table 1. Polynomials of the best individuals in generation 0, 40, 80, 120, 160, and 200 in the run of Fig. 3.

$t = 0$ $1.644a + c$

40 $1.05316a + 0.552276b + c$

80 $1.05316a + 0.199359b + 0.71841ab + c$

120 $0.0668757 + 1.08918a + 0.24139b + 0.954167c + 0.98242ab$
$-0.0551888bc$

160 $0.0643533 + 1.0481a + 0.251608b + 0.955896c + 1.00846ab$
$-0.00732622ac - 0.079142bc$

200 $0.0643533 + 1.0481a + 0.28893b + 0.955896c + 1.00846ab$
$-0.00658686ac - 0.137102bc - 0.00250218abc$

better). A large improvement occurs when computation in terms of particles and domains is discovered right around generation 80.

The further improvement towards the end of the run (approximately from generation 140) is an adaptation to the particular lattice size used. As shown in Fig. 5, the non-trivial dynamics of the boundary between a black region (with state values close to -1) and a checkerboard region causes it to change its propagation direction after around 50 time steps. This improves performance on this lattice size, but makes the generalization to other lattice sizes significantly worse. However, this adaptation provides an interesting example of behavior that depends crucially on the continuous degrees of freedom, and which could not be obtained in a cellular automaton.

Results from a larger number of runs with $r = 1$ were in many cases qualitatively similar to that of Fig. 3, indicating that the limited radius gives rather few options for good classification behavior, at least with reasonably simple transition functions.

We also performed a number of runs with $r = 3$, where a larger diversity of solutions was found. In that case, it was apparent that in the polynomial representation there are a number of algorithmically simple rules which perform reasonably well on the density classification task. In particular, a number of rules where the polynomial is a simple weighted linear sum of some of the states in the neighborhood (e.g, $\phi = \tanh(\beta(a + b + e))$) can give very good performance on easy training sets where densities cover a large fraction of the unit interval. In many cases, rules of this general form appear early in the evolutionary process, and with their close relatives form a quasi-stationary state which may persist for a long time.

The best $r = 1$ CML rule we discovered was found in a run without crossover which avoided this type of local minimum, and needed only 15 generations to discover the rule shown as a tree in Fig. 2. The constant β was set to 2.0 in this

Fig. 5. A simulation of the best rule from generation 200 in the run shown in Fig. 3

case. The tree in Fig. 2 corresponds to the polynomial

$$f(a, b, c) = a + c + 0.24637a^2b + 1.003821753a^2bc - 0.03413499a^2b^2c.$$

Fig. 6 shows a spacetime pattern from a simulation of this rule. When testing this rule on a test set of $3 \cdot 10^4$ initial states of length 149 generated from an uncorrelated and unbiased distribution, we found a classification accuracy of 0.842. This is significantly better than the $r = 3$ CA rules discussed in [1]. The rule performs significantly better on states with a majority of -1; in the test above the classification accuracy was 0.908 for examples intended to map to -1, and 0.779 for +1 examples. In the test the CML was simulated for a time of $T = 6N$, where N is the lattice size, to get a fair comparison with $r = 3$ rules (even if the CAs are allowed to run for the same time as the $r = 1$ CML, the CML performance is better).

The behavior of this rule is clearly quite similar to a CA rule. In particular one finds a structure consisting of domains and particles/domain walls between them. The equation $\phi(a, a, a) = a$ has two solutions corresponding to homogenous domains with $a = 0.999995$ or $a = -0.985411$. There is also a checkerboard domain where the local state values -0.999954 and 0.926398 alternate. These domains and their domain boundaries are clearly visible in Fig. 6. This suggests that a computational mechanics analysis in terms of domains and particles may be useful for this CML. *A priori* one might have imagined the existence of other computational mechanisms which made more extensive use of the continuous degrees of freedom (and even if domain walls appear, they need not move with constant velocity, as we have seen above).

Three regular domains are apparent for this rule: Checkerboard (C), White (W) and Black (B). The BW interface is stationary, and the CW interface moves with velocity -1 in CA-like fashion. The WC and CB interfaces are more interesting. The WC interface makes use of the continuous degrees of freedom to create an internal period of 4, which allows it to move with velocity -1/2. The CB interface is approximately periodic with period 7, and moves with velocity 3/7 (this should be checked more carefully on larger lattices). The results of collisions can be read off in Fig. 6.

In addition to these interfaces another class of particles can be identified, those which consist of single +1 in a background of -1, or vice versa. These particles could be regarded as WW and BB domain boundaries. An isolated WW (BB) particle moves with velocity -1 (+1) while it decays slowly and disappears. This is a feature of the rule that exploits the continuous values and cannot be reproduced by a CA. Several WW particles together form a checkerboard wedge, which represents a "don't know"-state where additional information is required. The detailed computational mechanics of this rule will be discussed elsewhere, but we can note a few features. In particular, a WW particle can only collide with a stationary BW boundary. When this happens, the BW boundary is moved one position to the right (and vice versa for a BB particle). In this way the location of the BW boundary keeps track of the balance between densities on its two sides.

The BW boundary is in turn annihilated either by a CB boundary from the left or a WC boundary from the right. The first case results in a spreading W region, the second in a spreading B region. In Fig. 6, the wedge of black just barely misses its own checkerboard tail as it sweeps around the edge of the lattice. In this way, the W region created by the CB-BW collision can take over the lattice and produce a correct classification.

4   Conclusions and Discussion

We have used genetic programming to evolve coupled map lattices for density classification with threshold 1/2. A solution with $r = 1$ was found which shows a significantly higher classification accuracy than previously discovered $r = 3$ cellular automata for the same task. The continuous degrees of freedom allow dynamical behavior that cannot be realized in a cellular automaton.

These results can still be improved in several ways. Our results for $r = 1$ could probably be improved with more computational effort. Other representations should also be explored. An alternative would be to use a representation more similar to a CA rule table, where the transition function ϕ is represented in terms of basis functions that are local in the space of neighborhood configurations. Similar ideas of local representations can be found, e.g, in radial basis neural networks, in kernel methods for density estimation, and in the wavelet transform. In this way, a single mutation only affects the transition function in a limited

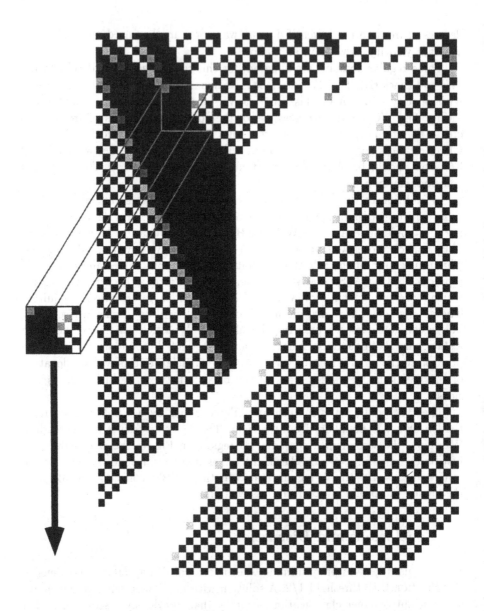

0.3879	-0.9854	-0.9854	-0.9845	0.9809	0.9999	-0.4481
-0.9997	-0.7464	-0.9854	-0.9844	0.9817	0.6064	0.9930
-0.9848	-0.9929	-0.9892	-0.9844	0.6819	0.9999	-0.8633
-0.9854	-0.9850	-0.9850	-0.9852	0.9292	-0.7221	0.9434
-0.9854	-0.9854	-0.9854	-0.9844	-0.9995	0.9770	-0.9995
-0.9854	-0.9854	-0.9854	-0.9854	-0.9854	-0.9999	0.9054

Fig. 6. An example of a space-time diagram of the best evolved rule. Numerical values are shown for a small subregion of the diagram.

region of the space of neighborhood configurations, in analogy with mutations on a CA rule table. Recombination can also be introduced in a way that mimics recombination on CA rule tables.

The use of continuous functions also gives us the possibility of using the error function on the training set for doing local minimization, e.g., by gradient descent like in the back-propagation algorithm.

The use of real-valued degrees of freedom also allows natural encodings of many other problems. Some simple variants of the present problem would be to consider other thresholds, to allow arbitrary input values, or to consider a continuous output giving the initial fraction of +1. More interesting and difficult problems could for example include calculating the median or the second largest element of the input. Coupled map lattices could also be evolved for more applied information processing tasks, e.g., in image processing.

References

1. Andre, D., Bennett III, F. H., and Koza, J. R.: Discovery by Genetic Programming of a Cellular Automata Rule that is Better than Any Known Rule for the Majority Classification Problem, in J. R. Koza, D. E. Goldberg, D. B. Fogel, R. L. Riolo, eds., *Genetic Programming 1996: Proceedings of the First Annual Conference*, 3–11 (MIT Press, 1996).
2. Capcarrere, M. S., Sipper, M., and Tomassini, M.: Two-state, r=1 cellular automaton that classifies density, *Physical Review Letters*, **77** (1996) 4969–4971.
3. Das, R.: The Evolution of Emergent Computation in Cellular Automata, Ph. D. thesis, Colorado State University, 1996.
4. Davis, L.: private communication cited in [1].
5. Gacs, P., Kurdyumov, G. L., and Levin, L. A., One-dimensional uniform arrays that wash out finite islands, *Problemy Peredachi Informatsii* **14** (1978) 92–98.
6. Holden, A. V., Tucker, J. V., Zhang, H., and Poole, M. J.: Coupled map lattices as computational systems, *CHAOS* **2** (1992) 367–376.
7. Kaneko, K.: *Physica D*, **34** (1989) 1.
8. Koza, J. R.: *Genetic Programming: On the Programming of Computers by Means of Natural Selection.*, (MIT Press, 1992).
9. Land, M. and Belew, R. K.: No Perfect Two-State Cellular Automata for Density Classification Exists, *Physical Review Letters*, **74** (1995) 1548–1550.
10. Lindgren, K.: Evolutionary Phenomena in Simple Dynamics, in *Artificial Life II*, C. G. Langton, C. Taylor, J. D. Farmer, S. Rasmussen, eds., 295–312 (Addison-Wesley, 1992).
11. Lindgren, K. and Nordahl, M. G.: Universal Computation in a Simple One-Dimensional Cellular Automaton, *Complex Systems*, **4** (1990) 299–318.
12. Mitchell, M., Hraber, P. T., and J. P. Crutchfield, J. P.: Revisiting the Edge of Chaos: Evolving Cellular Automata to Perform Computations, *Complex Systems*, **7** (1993) 89–130.
13. Mitchell, M., Crutchfield, J. P.,Hraber, P. T.: Evolving Cellular Automata to Perform Computations: Mechanisms and Impediments, *Physica D*, **75** (1994) 361–391.
14. Mitchell, M., Crutchfield, J. P., Das, R.: Evolving Cellular Automata to Perform Computations, in T. Bäck, D. Fogel, and Z. Michalewicz (Eds.), *Handbook of Evolutionary Computation*, (Oxford University Press).

15. Mitchell, M.: Computation in Cellular Automata: A Selected Review, SFI Working Paper 96-09-074, to appear in H. G. Schuster, and T. Gramms, (eds.), *Nonstandard Computation*, (VCH Verlagsgesellschaft, Weinheim).

16. Orponen, P. and Matamala, M.: Universal computation by finite two-dimensional coupled map lattices, *Proc. Physics and Computation 1996*, 243-247 (New England Complex Systems Institute, Cambridge, MA, 1996).

17. Packard, N. H.: Adaptation toward the edge of chaos, in *Dynamic Patterns in Complex Systems*, J. A. S. Kelso, A. J. Mandell, and M. F. Shlesinger, eds., 293–301 (World Scientific, 1988).

18. Sipper, M.:*Evolution of Parallel Cellular Machines*, (Springer-Verlag, 1997).

19. Wolfram, S. ed.:*Theory and Applications of Cellular Automata*, (World Scientific, 1986).

Genetic Programming for Automatic Design of Self-Adaptive Robots

Stéphane Calderoni and Pierre Marcenac

IREMIA - Université de La Reunion
15, Avenue René Cassin - BP 7151
97715 Saint-Denis messag cedex 9
tel: (+262) 93 82 87 - fax: (+262) 93 82 60
email: [calde][marcenac]@univ-reunion.fr
url: http://www.univ-reunion.fr/~calde/

Abstract. The general framework tackled in this paper is the automatic generation of intelligent collective behaviors using genetic programming and reinforcement learning. We define a behavior-based system relying on automatic design process using artificial evolution to synthesize high level behaviors for autonomous agents. Behavioral strategies are described by tree-based structures, and manipulated by genetic evolving processes. Each strategy is dynamically evaluated during simulation, and weighted by an adaptive value. This value is a quality factor that reflects the relevance of a strategy as a good solution for the learning task. It is computed using heterogeneous reinforcement techniques associating immediate and delayed reinforcements as dynamic progress estimators. This work has been tested upon a canonical experimentation framework: the foraging robots problem. Simulations have been conducted and have produced some promising results.

1. Introduction

In many research areas like robotics, computer programs have come to play an important role as scientific equipment. Computer simulations provide a powerful device for virtual experiments, as opposed to physical ones which are difficult to design, costly, or even worse dangerous. Although computer models provide many advantages over traditional experimental methods, some problems may be encountered too. In particular, the actual process of writing software is a complicated technical task with a high risk of error. Indeed, if scientists were constructing their own physical experimental equipment, nowadays researchers have to be programming wizard in addition to being scientists. This establishment has brought us to look into the design of a powerful platform in order to allow scientists to focus on their research rather than on building appropriate tools. This platform is intended to provide a complete software environment viewed as a virtual laboratory and providing agent-oriented support to model and simulate autonomous artificial systems like robots or animats.

Round this ambitious project, we have drawn several main lines of research. One of them consists in providing agents with self-adaptive capabilities in order to help them to move by their own in dynamic and non-deterministic environments. Self-adaptation is then considered as a behavior-based learning mechanism. All along this paper, we thus adopt a behavior-based vision in our agent design. A behavior-based system relies on coordination of some behavior modules, defined as primitives [Safiotti and al. 1995] [Dorigo and Schnepf 1993]. To design such a system thus consist in designing individual primitives first and then designing some kind of coordination mechanism to manage the interaction between them. However, increasing agent and task complexity can make the design difficult. Actually, in highly dynamic and in addition non-deterministic environment, such a system appears to be necessarily self-adaptive. Consequently, automatic design process using artificial evolution is advocated to synthesize high level behaviors for autonomous agents.

The main topic developed in this paper is the automatic generation of intelligent collective behaviors using coupling of genetic programming with reinforcement learning. We propose a real-time collective learning technique which makes agents behaviors emerging dynamically from genetic manipulations on their individual behavior descriptions. This learning mechanism ensures flexibility and adaptation to the disturbances of a highly dynamic environment during a simulation. Actually, the most important feature in evolutionary processes like genetic programming is basically the adaptation factor of genetic programs, since it reflects their relevance as good solutions for the problem to be solved. By another way, self-adaptation ability of a system is strongly correlated with learning from experiences over time. Our approach is thus to compute this adaptation factor with reinforcement learning techniques, based on experiences, by using two kinds of reinforcement indicators: immediate and delayed reinforcement.

In the first section, we present a brief overview of learning research in the domain of multi-agent systems, and especially reinforcement learning and genetic programming. This brings us to define, in the second section, our descriptive formalism of behaviors' representation, adapted to genetic handling techniques. Next, in the third section, we describe our collective learning algorithm coupling both genetic programming and dynamic reinforcement learning for real-time computing of adaptation factors. Finally, in the fourth section, we propose an experimentation framework to illustrate the learning approach applied on the foraging robots problem. Some results are then presented and discussed in conclusion of this work.

2. Learning in Multi-Agent Systems

[Weiß 1996] reports a wide variety of possible forms of learning in multi-agent systems according to standard criteria from machine learning field:

- the *learning strategy* used by a learning entity,
- and the *learning feedback*, available to a learning entity and which indicates the performance level achieved so far.

Our learning strategy is a kind of *learning by discovery*, gathering new knowledge and skills by making observations (sensing information from the environment), conducting experiments (acting according to a genetic program), and generating and testing hypotheses or theories on the basis of the observational and experimental results (evolving genetic programs). The learning feedback, obtained by reinforcement, specifies the utility of the actual activity of the learner and the objective is to maximize this utility. The two next subsections develop these two aspects of learning mechanism.

2.1. Reinforcement Learning

The most significant works on learning in multi-agent systems are probably those dealing with reinforcement learning. In this type of learning, the tendency to perform an action is reinforced if it has produced satisfactory results in the past, whereas it is weakened otherwise. In a general way, each action is associated with a reward or a punishment at a given time. The learning process maximizes rewards at long-range, that is to say maximizes its expected success. The Q-Learning algorithm [Watkins and Dayan 1992] estimates the reward for each action in long-term and constitutes a recursive mechanism of the choice of both actions to perform, and their schedule. The use of Q-Learning agents [Sandholm and Crites 1995] leads to preferential choices of the best actions, and to the surrender of actions whose chance of success is weaker. Reinforcement learning tasks are typically implemented using a monolithic reinforcement function that enables the learner to acquire the optimal policy. Constructing such a reinforcement function can be a complicated task in highly dynamic domains. Indeed, the environment may provide some heterogeneous feedback, as immediate rewards, or intermittent reinforcement and a handful or fewer of very delayed information associated with reaching one or more high-level goals. [Mataric 1996] discusses how such heterogeneous multi-modal reinforcement, based on progress estimators, can be combined and utilized to accelerate the learning process.

2.2. Genetic Algorithms and Genetic Programming

Another field of learning research is genetic algorithms and genetic programming which provide a revolutionary heuristic mechanism to make computers able to solve problems without explicitly programming the right method of resolution.

Genetic algorithms were first studied by J. Holland and al. in the University of Michigan. This preliminary work was synthesized in the 70's in [Holland 1975]. Genetic algorithms are stochastic algorithms of exploration and optimization which rely on mechanisms derived from natural and genetic evolution. They imitate the adaptive capabilities of living organisms in their natural environment, and the information exchange process used by our genetic material.

In the domain of machine learning, genetic algorithms seems to constitute a mechanism of exploration more human than classical techniques implemented in traditional computer systems. Indeed, they manipulate arbitrary structures, and speculate on these structures, by looking for the best alternatives and by combining intuitions. They are inductive, as opposed to deductive mechanisms of exploration in traditional artificial intelligence systems.

In general, a genetic algorithm simulates the evolution of a population of individuals. An individual is defined by his *genotype*, a data structure which embodies his properties. In most of the cases, this data structure is a vector of symbols. The population evolves stochastically in a discretized time. Each individual moves in an environment (the context defined by the tackled problem) in which he is more or less adapted. The measure of this adaptation degree is explicit and indicates his capability to produce clones of himself. Thus, his genotype is expressed by a function, called the adaptation function, in order to obtain his *phenotype*. The phenotype of an individual is the expression of his genotype within the simulated domain and indicates his capability to survive in his environment (and thus to reproduce himself). At each generation, a mechanism shapes the population of the next generation, by selecting the best individuals.

Applications of genetic algorithms are very diversified. In particular, there are two main research domains conducted in machine learning using genetic techniques: on the one hand, learning in knowledge-based systems, where the best known example is the learning classifier systems [Holland and Reitman 1978]; and on the other hand, the automatic program generation with genetic programming. This last research domain constitutes an actual interesting approach in our purpose of building synthetic complex behaviors, and is thus tackled in this paper.

Both reinforcement learning and genetic programming have been proven to be generally efficient in learning tasks. That is the reason why we decided to inherits from these two paradigms, and combine them in an extended way for collective learning in multi-agent simulation.

3. A Genetic Coding of Agent's Behavior

While genetic algorithms manipulate strings of symbols, genetic programming manipulates more complex structures like computer programs. J. Koza was one of the first researchers whom proposed this kind of approach [Koza 1989]. In genetic programming, the individuals which constitute a population are viewed as hierarchic compositions of functions and terminations (a constant or a variable), chosen in a finite set, determined according to the tackled problem. Search space thus corresponds to all the combinations which can be created on this set. Then, genetic programming have attracted a large number of researchers because of the wide range of applicability of this paradigm, and the easily interpretable form of the solutions that are produced by these algorithms [Kinnear 1994] [Koza 1994].

In the agent domain, a behavior-based system generally relies on coordination of some behavior modules, defined as primitives. The design of such a system consists in designing both primitives and the way in which they are coordinated. [Brooks 1986] proposed the *subsumption* architecture for reactive agent's control. The subsumption architecture decomposes an agent in vertical modules. Interactions between modules are fixed and performed under a dominance relationship defined at the design step. However, such a behavioral controller is too rigid, and is unable to evolve since the designer has definitively fixed the global behavior of the system when becoming operational. On the opposite, genetic programming does not hardwire a priority policy in advance, since genetic mechanisms make the relationships between primitives evolving over time.

The agent behavior is the dynamic policy with which he moves in his environment, combining sensing and acting primitives. In order to describe such a scheme of primitives, we propose a structural formalism based on a tree structure viewed as a parse tree. Furthermore, this formalism is conveniently adapted to genetic handling mechanisms. Actually, the parse tree is considered as the digital chromosome of the agent. It constitutes his genetic program encoding his global behavior. More precisely, the behavioral primitives are associated with nodes of this parse tree and organized upon the edges linking nodes. More specifically, we can distinguish two kinds of primitives: primitives which are associated with the sensors of the agent (module which make the agent perceiving informations from his environment), and primitives which are associated with his actuators (modules which enable him to make actions on his environment). Thus we respectively define two kinds of operators in order to embody each of these complementary primitives within the parse tree.

- The first operator is intended to perform a sensing-primitive through the sensors. This operator is a conditional operator since the sensing-primitive is assumed to return a boolean evaluation depending on the success of the primitive. We called it IfThenElse, and it runs as described in figure 1 below:

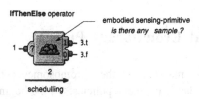

1. The operator takes the control.
2. The embodied sensing-primitive is performed on sensors and returns a boolean value.
3.t. If the sensing-primitive returns true, then the control is passed to the 1 successor.
3.f. Otherwise, the control is passed to the O successor.

Fig. 1. the IfThenElse operator mechanism

- The second operator is an operator intended to perform an action-primitive through actuators. This operator is an unconditional operator. We called it Do, and it runs as described in figure 2 below:

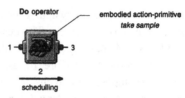

1. The operator takes the control
2. The embodied action-primitive is performed on actuators
3. The control is then passed to the successor in the parse tree

Fig. 2. the Do operator mechanism

These operators can then be arranged within a parse tree to express complex behaviors, combining, on the one hand, the primitives that sense information from the environment, and on the other hand, primitives that make the agent acting in the environment. The parse tree is skimed through by a controller that successively performs each primitive encountered in nodes. When the controller reaches a leave, it restarts from the root, authorizing the behavior to be performed in a loop during the agent life-cycle. By this way, according to the context of the environment, and then to the inquiries sensed by the agent, some edges can be skimed through several consecutive times as a local loop within the whole behavior. Indeed, such a mechanism can be assimilated to conditional loops such as while-loops, or repeat-until-loops.

This formalism provides an actual flexibility since it is fully reconfigurable thanks to its structure. Moreover, it is sufficiently expressive to describe a wide variety of high-level behaviors, since it authorizes control structures, whether they are conditional or not. In addition, it's extremely simple to interpret a behavior from a tree-based description. Our software platform provides a user graphical interface which increases the easiness of such an interpretation, with very expressive widgets. This property applies to users who want to study complex behaviors without having to decode some hundreds of esoteric code lines !

4. The Behavioral Evolutionary Process

The evolutionary process manipulates the chromosomes defined in the previous section, according to the classical genetic policies. At the beginning of simulation, a random generation of initial chromosomes is generated in so far as agent behavioral strategies. We could say that such an encoded behavioral strategy corresponds to an innate behavior, in opposition with the learned behavior which will be derived during simulation and which constitutes an acquired feature. Then, when a behavioral chromosome has been assigned to each agent, the evolving process can be split into three successive main stages, as follows:

1. evaluation of the behavior's relevance with adaptation factor for applicant chromosomes,

2. selection of the best strategies, according to adaptation factors,
3. reproduction of these best strategies and surrender of weaker ones.

This sequence constitutes a genetic cycle within which a generation of behavioral chromosomes is evaluated. At the end of such a sequence, a new generation appears by reproduction of the selected individuals from the previous generation, and a new cycle is started with new applicants. The period of each genetic cycle can be set during the simulation.

4.1. Evaluation Process

During this stage, each agent is testing his genotype by decoding his behavioral chromosome, and performing primitives on his environment. This experimentation process is assumed to provide an evaluation of the expression of the behavioral strategy, that is to say a quality factor of the phenotype. This quality factor denotes the level of adaptation of the strategy which controls the attitude of the agent in his environment. By this way, this factor should reflects the relevance of the agent behavior. Anyway, this kind of evaluation has to be dynamically computed. Indeed, each action-primitive leaves some tracks in the environment, and contributes to approximate or wander from the ultimate goal of the sought after the learning task. Thus, it would be difficult to globally evaluate the whole behavior at the end of the evaluation process by establishing wether the global task has been learned.

Indeed, each action-primitive should induce a dynamic local evaluation, and then be combined into a global factor to balance the whole behavior. This kind of representation of local evaluation connects the reinforcement learning technique. Furthermore, each local reinforcement may be of two complementary models: on the one hand, it can be immediate, and on the other hand, it can be put off. Immediate reinforcement is applied as soon as an action-primitive has been performed, whereas delayed reinforcement supposes that an observer mechanism is triggered to control the behavior evolution at long term.

In order to install such a reinforcement mechanism, we based our work on those described in [Mataric 1997a] who uses such a reinforcement technique in a learning process of action's selection. Her learning system is table-based, and maintains a matrix whose entries reflect a normalized sum of the reinforcement received for condition-action pairs over time. The values in this matrix fluctuate according to the received reinforcement. They are collected during the execution of an action and updated and normalized when actions are switched.

Let R the reinforcement function which dynamically balances the adaptation factor of an individual. We define R as follows:

$$R = \sum_t \left[\alpha \cdot r_i(t) + \beta \cdot r_d(t) \right] \text{ with } \alpha, \beta \geq 0 \text{ and } (\alpha + \beta) = 1$$

where $r_i(t)$ represents the immediate reinforcement received over time, whereas $r_d(t)$ represents the delayed reinforcement. Moreover, both $r_i(t)$ and $r_d(t)$ can either represent a positive reinforcement, or a negative one. The global learning function that balances the above reinforcements simply sum them over time. The influence of the different types of feedback can be weighted by introducing two weight factors α and β which measure their contribution to the sum.

4.2. Selection Process

At the time of the selection, the system is assumed to hold the best strategies according to the quality factor defined previously. This principle leads to make the most adapted behaviors surviving and allowing them to be inherited at the next generation. Several selection methods have been proposed in the litterature. We have chosen the rank method, described in [Baker 1985], because of its ability to not prematurely privilege a good individual, and then to preserve the whole population of being rapidly dominated by a super-individual. Indeed, rather than fix the number of descendants of individuals proportionately to their performance, this number is limited between min and max boundaries, and computed according to the rank of individuals, sorted by their performance.

4.3. Reproduction Process

After the selection process, the evolving mechanism is then trusted to the reproduction process which relies on two fundamental mechanisms of genetic evolving processes: crossing-over and mutation.

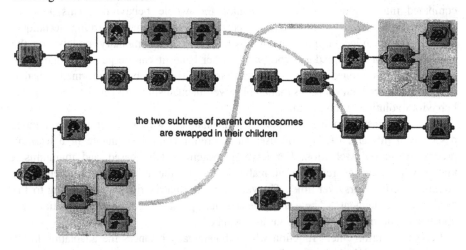

the two subtrees of parent chromosomes
are swapped in their children

Fig. 3. the crossing-over mechanism

The crossing-over is a mechanism which, from two initial chromosomes, produces two new chromosomes that have inherited some local features from their parents thanks to the exchange of some local properties. The two new chomosomes then replace their parents in the current population. In the case of our chromosome formalism, the crossing-over operates upon a pair of parse trees by randomly choosing a subtree within each parse tree, and by swapping them, as shown in the figure 3 above. The new genetic programs are supposed to contain good subprograms from their parents, combined in a new way.

Finally, the mutation mechanism is playing a secondary role in comparison with the crossing-over, but it should not be neglected. Indeed, it avoids a premature convergence of individuals by keeping a diversity within the whole population. For that, an operator is randomly chosen within the parse tree, and its primitive is replaced with another, also randomly chosen.

Crossing-over and mutation are not systematically applied, but rather in accordance with some probabilistic rates, set with the simulator. These two factors represent some important features of a genetic algorithm, since its efficiency depends on their respective values. In practice, a significant probability is set for the crossing-over rate (about 0.6), in comparison with the mutation one which is generally inversely proportional with the size of population (0.1 for a population of 10 individuals). [De Jong 1975] presents a study of the choice of these factors, upon the function optimization problem. Furthermore, the size of the population should be moderate.

5. Application to the Foraging Robots Problem

Since [Steels 1989], in the domain of collective problem solving, a large part of research is devoted to applications which bring exploratory robots into play. The most common problem is probably the foraging robots problem. In this case study, the goal is to make a team of robots find and collect ore samples in an unpredictable environment and take them back to a home base.

This problem framework seems to be successful as it is used in a lot of possible derived applications in artificial intelligence and artificial life. Indeed, it serves as a canonical abstraction of a variety of real-world applications. It tackles the general problem of finding resources (e.g., food for animals, or samples for robots) in an unknown environment, and bringing them to a specified location (e.g., the nest for insects, or the home base for robots). This collective learning task is complex and submitted to high dynamicity. We can quote the works conducted in ethology by [Deneubourg and al. 1986], in exploration of planets or dangerous locations by [Brooks 1990], and in genetic programming by [Koza 1990] with the *Santa-Fe Trail*. [Drogoul and Ferber 1993] presents different implementations of *Tom Thumb Robots*, whose behavior is based on the foraging behavior of ants. More recently, [Mataric 1996] proposes reinforcement learning algorithms for fully autonomous robots. This work is extended in [Mataric 1997b] by the introduction of social criteria into the learning algorithms, in order to improve the collective aspects of the learning task.

5.1. The Experimental Framework

The key idea is to make a set of little robots collect objects (ore samples) in an unknown environment. This environment is formalized by a $n \times n$ grid-world with a toroidal structure. It contains a variable population of robots, a home base, and samples of ore spread in the environment. Robot movement is limited to either a vertical, horizontal or diagonal step per cycle. Two agents are allowed to occupy the same location at the same time. They have a limited visual field of depth d, defined according to the Manhattan distance. They accurately locate the relative position and recognize the type of any object (or other agent) standing within their visual field. Figure 4 below shows the artificial world modeled as the experimental environment to validate our purposes.

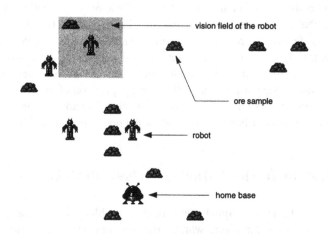

Fig. 4. the foraging robots problem

Each agent has a finite repertoire of simple primitive behaviors:

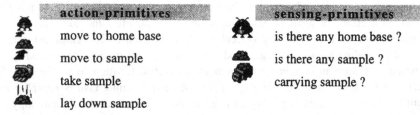

action-primitives	sensing-primitives
move to home base	is there any home base ?
move to sample	is there any sample ?
take sample	carrying sample ?
lay down sample	

For the robots, the learning task consists in finding a skillful combination of these primitives in order to build advanced behaviors that allow them to find and bring back a maximum of samples.

The behavioral strategies built under these primitives are dynamically evaluated through reinforcement process whose factors are the following:

The immediate reinforcement is defined as follows:

- $r_i(t) = p_s > 0$ when a sample is picked up. This reward favors programs which find samples and pick them up;
- $r_i(t) = l_s > 0$ when a sample is laid down at the home base. This one favors programs that reached to bring samples back to the home base;

- $r_i(t) = b_p < 0$ when a sample is not picked up even though the agent occupies the same location. This punishment penalizes programs which do not pick up a detected sample;
- $r_i(t) = b_s < 0$ when a sample is laid down away from home base. This one penalizes programs which scatter the samples anywhere in the environment;

p_s, l_s, b_p and b_s are factors set by the user during the simulation.

The delayed reinforcement $r_d(t)$ is a progress estimating function associated with homing, and initiated whenever a sample is picked up. If the distance to home base is decreased while $r_d(t)$ is active, the agent receives positive reinforcement. If the agent keeps motionless, no reinforcement is delivered. Finally, movement away from home base is punished by a negative reinforcement. $r_d(t)$ is disactivated when the home base is reached.

5.2. Results

The simulations were conducted on a prototype which we are currently developing in Java language: the SALSAA[1] platform. This is an agent-oriented platform for modeling and simulation of complex adaptive systems.

First, the simulations consisted of generating some random behaviors for each robot, by controlling the maximum depth of parse trees. Then, we let these behaviors evolve under genetic programming mechanisms. All the simulation's factors were set with the simulator. The following have been retained to illustrate our purposes:

Domain's factors		Genetic factors	
environment's grid size	20x20	period of genetic cycles	100 time units
robots' population size	20	crossing-over rate	0.6
total amount of samples	60	mutation rate	0.1

Reinforcement factors	
p_s	2
l_s	100
b_p	-1
b_s	-1
α	0.5
β	0.5

In a second step, we have simulated a hand-coded behavior, without any genetic handling, in order to compare the two approaches, and to evaluate the relevance of emerging behaviors. Figure 5 below describes the hand-coded behavior embodied in each robot.

---

[1] SALSAA is an acronym for Simulation of Artificial Life with Self-Adaptive Agents

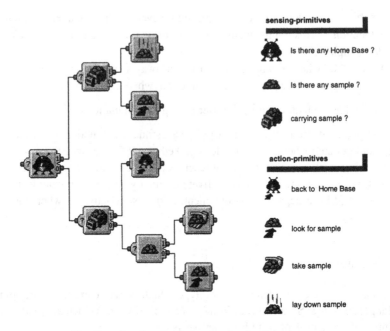

Fig. 5. a hand-coded behavior for the foraging task

In each simulation, the amount of samples brought back to the home base has been reported as a function of time on curves, like those represented by the figure 6, in order to measure the global performance of the population over time.

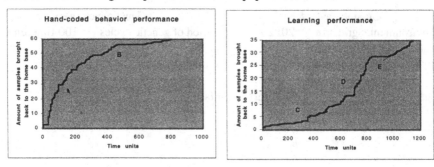

Fig. 6. performance curves of hand-coded behavior vs learned behaviors

Furthermore, some interesting global phenomena become apparent. Indeed, the first curve, which corresponds to the hand-coded behavior, presents a sheer gradient in its first part (A), then the amount of brought back samples progressively decreases over time (B). This phenomenon is due to the gradual decreasing of the density of samples in the home base's neighborhood. As and when robots bring back samples, they become rarefied near the home base location. So, robots need more and more time to find again other samples.

In the second curve, the random behaviors generated at the beginning of the simulation labouriously achieve the foraging task (C). When the first robots discovered a good way to perform this task, the genetic mechanisms propagate some good sub-behaviors throughout the whole population. Progressively, the global performance of the population increases, due to the dominance of good emerging behaviors (D). Then, the depletion of samples should decrease this performance (E) because of the increasing distance between samples and the home base. So, the current good behaviors are progressively punished, implying a reconfiguration of the genetic programs, which finally arrive to make another solution emerging, then a similar process continues the phenomenon. This shows the robustness of such automatic design of complex behaviors. Indeed, the robots dynamically adapts their behavior to the unpredictable context of their environment.

Furthermore, our approach seems to be promising when we glance at the progression' speed of the learning process. Indeed, at the beginning of the simulations, all behaviors are generated randomly. So robots do not know how to accomplish the foraging task. Then, progressively, the population's efficiency becomes exponential (C-D) (until samples become rarefied).

Finally, one of the most interesting results provided by our platform is the explicit description of the best emerged behavior. And, as shown in figure 7 below, it is not necessarily the most deliberate behavior that is the most relevant.

Fig. 7. one of the best emerged behaviors

The above behavior consist in taking any sample within the perception field, then in moving of one step towards the home base, then in laying down the eventual carried sample, and finally in moving of one step towards any localized sample. Obviously, the robot stands on the current cell, near the sample which has been laid down by the previous primitive. And then, the same sequence is reproduced: he takes the sample, and move towards the home base, and so on. It is clear that this solution is not the most efficient, but it works, and it has been totally discovered by the robots.

6. Conclusion and Further Work

This paper has both presented a formalism to describe agents behaviors and a collective learning algorithm addressed to autonomous agents in dynamic and unpredictable environment.

We have introduced a very simple formalism for the description of agents behaviors, based on a tree structure and two kinds of operator: one for action-primitives, and a second one for sensing-primitives. By combination, they authorizes the expression of high complex behaviors, with both conditional and unconditional control structures.

By another way, this formalism is fully adapted to genetic programming. Thus it provides to the agents a powerful mechanism of self adaptation by producing dynamic and reconfigurable complex behaviors from simple primitives. Furthermore, the emerging process has been proven robust, and capable of adaptation in an unpredictable environment.

Finally, it has the advantage of providing an explicit and easily interpretable form of agents behaviors.

Nevertheless, we have to point out some failings in this approach. The reinforcement function evaluates the whole behavior, and this is this evaluation that is used to select the applicants to reproduction. The reproduction essentially involves the crossing-over mechanism. This mechanism cutts off parts of behaviors and mixes them up into new behaviors. Now, a behavior relies on the order in which its primitives are performed. Thus, a blind crossing-over may destroy a very good behavior by separating a fundamental sequence of primitives. An alternative to this problem may be a local evaluation of behaviors by weighting the edges of the tree. Indeed, the weight of edges could express their resistance to the cutting-off, and progressively protect the adjacent primitives which are relevant when they are performed successively. Such an alternative will be tackled in further works.

At long term, we aim to propose to scientists like ethologists or roboticists a complete environment of modeling and simulation with agent-oriented support. They will be able to design easily their model with a fully interactive programming environment based on graphic programming language. It will allow them to program their autonomous systems in an easily interpretable behavior-based language, and to make them evolve dynamically with genetic techniques in order to identify complex behaviors totally learned, without explicit programmed solutions.

The progress of the SALSAA project will be detailed at the following url:

http://www.univ-reunion.fr/~calde/salsaa/

References

[Baker 1985] J.E. Baker, "Adaptive Selection Methods for Genetic Algorithms", in *Proceedings of the First International Conference on Genetic Algorithm*, pp. 101-111, Eds. L. Erlbaum Associates: Hillsdale, Pittsburgh, USA, 1985.

[Brooks 1986] R.A. Brooks, "A Robust Layered Control System for a Mobile Robot: An Approach to Programming Behavior", in *IEEE Journal of Robots and Automation*, Vol. RA-2(1), pp. 14-23, 1986.

[Brooks 1990] R.A. Brooks, "Elephants Don't Play Chess", in *Journal of Robotics and Autonomous Systems*, Volume 6, pp. 3-15, 1990.

[De Jong 1975] K.A. De Jong, "An Analysis of the Behavior of a Class of Genetic Adaptive Systems", *PhD thesis*, University of Michigan, USA, 1975.

[Deneubourg and al. 1986] J.L. Deneubourg, S. Aron, S. Goss, J.M. Pasteels and G. Duerinck, "Random Behaviour, Amplification Processes and Number of Participants: How they Contribute to the Foraging Properties of Ants", in *Physica 22D*, pp. 176-186, Amsterdam, North Holland, 1986.

[Dorigo and Schnepf 1993] M. Dorigo and U. Schnepf, "Genetics-based Machine Learning and Behaviour-based Robotics: A New Synthesis", in *IEEE Transactions on Systems, Man and Cybernetics*, 23, 1, pp. 141-154, 1993.

[Drogoul and Ferber 1993] A. Drogoul and J. Ferber, "From Tom Thumb to the Dockers: Some Experiments with Foraging Robots", in *From Animals to Animats II*, pp. 451-459, MIT Press, Cambridge, USA, 1993.

[Holland 1975] J.H. Holland, "Adaptation in Natural and Artificial Systems", Michigan Press University, Ann Arbor, USA, 1975.

[Holland and Reitman 1978] J.H. Holland and J.S. Reitman, "Cognitive System based on Adaptive Algorithms", in *Pattern Directed Inference Systems*, Eds. D.A. Waterman and F. Hayes-Roth, Academic Press, New-York, USA, 1978.

[Kinnear 1994] K.E. Kinnear, Jr., editor, "Advances in Genetic Programming", MIT Press, Cambridge, USA, 1994.

[Koza 1989] J.R. Koza, "Hierarchical Genetic Algorithm Operating on Populations of Computer Programs", in *Proceedings of the 11th International Joint Conference on Artificial Intelligence*, Eds. Morgan and Kaufman, San Mateo, USA, 1989.

[Koza 1990] J.R. Koza, "Evolution and Co-Evolution of Computer Programs to Control Independent-Acting Agents", in *From Animals to Animats*, MIT Press, Cambridge, USA, 1990.

[Koza 1994] J.R. Koza, "Genetic Programming II, Automatic Discovery of Reusable Programs", MIT Press, Cambridge, USA, 1994.

[Mataric 1996] M. Mataric, "Learning in Multi-Robot Systems", in *Adaptation and Learning in Multi-Agent Systems*, Eds. G. Weiß and S. Sen, *Lecture Notes in Artificial Intelligence*, Vol. 1042, pp. 152-163, Springer-Verlag, 1996.

[Mataric 1997a] M. Mataric, "Reinforcement Learning in the Multi-Robot Domain", in *Autonomous Robots*, Vol. 4, N°1, pp. 73-83, January, 1997.

[Mataric 1997b] M. Mataric, "Learning Social Behaviors", to be published in *Journal of Robotics and Autonomous Systems*, 1997.

[Safiotti and al. 1995] A. Safiotti, K. Konolige and E.H. Ruspini, "A Multivalued-logic Approach to Integrating Planning and Control", in *Artificial Intelligence*, 76, 1-2, pp. 481-526, 1995.

[Sandholm and Crites 1995] T.W. Sandholm and R.H. Crites, "On Multi-Agent Q-Learning in a Semi-Competitive Domain", in Proceedings of the 14th International Joint Conference on Artificial Intelligence, Workshop on Adaptation and Learning in Multi-Agent Systems, pp. 191-205, Montreal, Canada, 1995.

[Steels 1989] L. Steels, "Cooperation between distributed agents through self-organization", in *Journal of Robotics and Autonomous Systems*, Amsterdam, North Holland, 1989.

[Watkins and Dayan 1992] C. Watkins and P. Dayan, "Technical Note on Q-Learning", in *Machine Learning*, Vol. 8, pp. 279-292, Kluwer Academic Publishers, Boston, USA, 1992.

[Weiß 1996] G. Weiß, "Adaptation and Learning in Multi-Agent Systems: Some Remarks and a Bibliography", in *Adaptation and Learning in Multi-Agent Systems*, Eds. G. Weiß and S. Sen, *Lecture Notes of Artificial Intelligence*, Vol. 1042, pp. 1-21, Springer-Verlag, 1996.

Genetic Modelling of Customer Retention

A.E. Eiben[1] and A.E. Koudijs[2] and F. Slisser[3]

[1] Dept. of Comp. Sci., Leiden University, The Netherlands
gusz@wi.leidenuniv.nl
[2] Cap Gemini, Adaptive Systems bv, The Netherlands
AKoudijs@inetgate.capgemini.nl
[3] Strategic Management & Marketing, University of Amsterdam, The Netherlands
slisserf@james.fee.uva.nl

Abstract. This paper contains results of a research project aiming at the application and evaluation of modern data analysis techniques in the field of marketing. The investigated techniques are: genetic programming, rough data analysis, CHAID and logistic regression analysis. All four techniques are applied independently to the problem of customer retention modelling, using a database of a financial company. Models created by these techniques are used to gain insights into factors influencing customer behaviour and to make predictions on ending the relationship with the company in question. Comparing the predictive power of the obtained models shows that the genetic technology offers the highest performance.

1 Introduction

Evolutionary Algorithms, in particular genetic programming, are relatively new and powerful techniques for analysing data. Their success in data analysis applications is, however, somewhat hidden. Quite a few application oriented research projects are classified as "confidential" - precisely by being successful and thus representing a competitive advantage. In this paper we report on a comparative study in a financial environment. The paper is meant as a documentation of a typical research project in the area illustrating various aspects, rather than a technical report focussing on optimal crossover rates and the like.

An application of evolutionary techniques in the field of direct marketing was a part of a research programme Intelligent Database Marketing, coordinated by the Foundation Marketing Intelligence & Technology. In particular, genetic programming was compared to two classical data analysis techniques used in marketing, logistic regression and CHAID (CHi-square Automatic Interaction Detection), furthermore a novel technique, Rough Data Analysis, was tested. The research has been carried out in cooperation with a Dutch mutual fund investment company that provided us with part of their database containing information about more then 500.000 clients. One of the main objectives of the mutual fund investment company is to increase their cash flow. One way to achieve this is to reduce the cash outflow by preventing clients to quit their relationship with the company. A climbing defection rate is a sure predictor of a diminishing flow of cash from customers to the company - even if the company replaces the lost customers - because older customers tend to produce greater cash flow and profits. They are less sensitive to price, bring in new customers and do not have acquisition or start-up cost. In some industries, reducing customer defections by as little as five percent points can double profits [7]. Customer retention is therfore an important issue. To be able to increase customer retention the company has to be able to predict which clients have a greater probability of defecting. This makes it possible for the company to focus their actions on the most risky clients which is necessary to make their efforts profitable. To keep clients from going it is also necessary to know what distinguishes a stopper from a non stoppers especially with respect to characteristics which can be influenced by the company.

2 Research methodology

Our research was carried out in different phases, similarily to earlier projects, [1]:

1. defining the problem and designing conceptual models with particular attention to relevant variables;
2. acquiring and arranging data, creating 3 distinct data sets for respectively training, testing and validating models;
3. analysis of data by four techniques independently;
4. interpretation of the four obtained models, and comparison of model performance.

The use of three different data sets was motivated by our willing to avoid overfitting of the models as well as for the sake of a fair comparison between the techniques. The training and test set were made avialable to each of the four research groups. Every group developed different models using only the data in the training set and selected one model by comparing them on the test set. This way we could prevent overfitting, by ruling out models that performed good on the training set but showed worse performance on the test set. The validation set was kept confidential, i.e. none of the four groups had access to it before proposing their best model for a final comparison.

Instead of a detailed description of the applied techniques we provide here references to the literature. Genetic programming is well described in [4], and further details on the genetic procedure used here are given in [8]. The CHAID (CHi-square Automatic Interaction Detection) technique is described in [2, 6] and see [3] for logistic regression. Rough Data Analysis is a novel technique, and its first description can be found in [5]. In this paper we only provide a detailed description of the evolutionary project, details of the work of other groups are subject of forthcoming reports.

3 Problem and data description

The research problem could be defined as follows:

What are the distinguishing (behavioural) variables between investors
that ended their relationship with the company (stoppers) from investors
that continued (non stoppers) and how well can different techniques/models
predict that investors will stop in the next month.

The dataset we finally defined and got consisted of 213 variables and 14394
cases in total, which was split in sets for training, testing and validating models.
The exact number and percentages of stoppers and non stoppers are given in
Table 1.

	stoppers	non stoppers	total
training	4006 (39.6%)	6109 (60.4%)	10115
test	837 (39.0%)	1309 (61.0%)	2146
validation	823 (38.6%)	1310 (61.4%)	2133
total	4006 (39.4%)	6109 (60.6%)	14394

Table 1. Specification of the data sets

Each case in the dataset can be seen as a investor on a specific moment in
time. If an investor had stopped the stopmonth was known. These stopmonths
were in the range of january 1994 till february 1995. For the non stoppers a
month in this range was randomly selected. Then for every investor timeseries
for 9 variable types were computed on a monthly bases, starting in the month
before the (stop)month. Each timeserie consisted of values of the 23 months
before the (stop)month with respect to the specific variable type. The variable
types were for instance:

– number of funds invested in,
– amount of invested money, and
– portfolio risk.

Additionaly the state of five profile variables was calculated in the month before
stopping. These state variables were:

– age,
– the period the (investment) relation existed,
– the invested capital two years before the stopmonth, and
– the (stop)month.

The binary output variable *is–stop* was 1 if an investor had no more money
invested with the company. Otherwise it was a 0. An investors stopmonth was
considered to be the first month during the period january 1994 till february
1995 that this state occured.

A final remark with respect to the data is that every investor analysed had at least a history of two years with the company. This is the case for about 90% of the investors.

4 Evolutionary Modelling

The evolutionary working group analysed the data and created prediction models by genetic programming, in particular by using the system OMEGA, provided by Cap Gemini, Holland. The OMEGA system is mainly deployed in the areas of marketing, creditscoring, stock prediction and symbolic regression.

OMEGA is a genetic programming system that builds prediction models in two phases. In the first phase a data analysis is performed, during which re-coding is carried out to transform the values of the variables to statistically significant different intervals. These recoding results can also be used within OMEGA to generate straightforward linear regression and CHAID models. Also during this phase, an exhaustive search is performed to find those bivariate com-binations (i.e. pairs) of variables that had the highest predictive power. In the second phase the genetic modelling engine initialises with a targeted first gener-ation of models making use of the bivariate combinations obtained in the first phase. After initialisation, further multivariate optimisation then takes place by using a genetic programming scheme which has its control parameters dynami-cally adjusted for consecutive generations. It is found that in practical situations OMEGA's genetic programming approach after a limited number of generations always outperformes the linear regression and CHAID models based on the same recoding of the variables, both in measures of accuracy and generalisation.

In modelling applications the most frequently applied measure for evaluating the quality of models is 'accuracy', which is the % of cases where the model cor-rectly fits. When building models for dichotomous behaviour, the so-called CoC measure is a better option for measuring model quality. The CoC (Coefficience of Concordance) actually measures the distinctive power of the model, i.e. its ability to separate the two classes of cases, see [8] for the definition. Using the CoC prevents failures caused by accepting opportunistic models. For instance, if 90 % of the cases to be classified belongs to class A and 10 % to class B, a model simply classifying each case as A would score 90 % accuracy. When the CoC is used, this cannot occur. Therefore, we used the the CoC value of the models as fitness function to create good models. Let us note that selection of the best model happened by measuring accuracy (on the test set), and that the final comparison between the four different techniques was also based on the accuracy of the models. Yet, we decided to use the CoC for evolving models in order to prevent overfitting. Control experiments (not documented here) con-firmed that evolving models with a fitness based on accuracy results in inferior performance.

As stated previously, OMEGA builds prediction models in two phases. The first phase, a statistical data analysis, indicates those variables that are of interest out of all supplied variables. Of the most interesting variables a summary of their individual predictive performance measured in CoC is shown in Table 2.

amount invested	71.1
risk of portfolio	70.7
s-ver	64.9
nr. of funds	62.7
duration of investment	62.0
duration of relation	60.4

Table 2. Results of data analysis: CoC value of the 6 most powerful variables

By the data analysis carried out in the first phase OMEGA is able to create a good initial population by biasing the chances of variables to be included in a tree. The initial models were generated into 2 populations of 20 trees each. During this initialisation, the 40 models varied in performance from 72.3 till 79.9 measured in CoC. Computing the accuracy of the best model on the training set gave a value of 73%. The relatively good CoC values after initialization show an increased joint performance compared to the best individual variable performance (which was 71.1 for *amount invested*). This is due to the use of special operators that act on the variables and the optimised interactions between the selected variables in the models. Till sofar, no genetic optimisation has taken place and several satisfactory models have been found.

During the genetic search we were using 0.5 crossover rate, 0.9 mutation rate and 0 migration rate between the two sub-populations, that is no migration took place. The two populations were only used to maintain a higher level of diversity. The maximum number of generations was set to 2000. After the genetic search process the best performing tree had a CoC value of 80.8 and a corresponding accuracy of 75% on the training set. For this particular problem the genetic optimisation phase does not show a dramatic improvement in performance with respect to the initial population. Note, however, that in the given application area each percent of improvement can imply considerable cost reductions or profit increase.

5 Analysis of the results

In this section we evaluate the outcomes from two perspectives. Firstly, we observe the best model and draw conclusions on customer behavior. Here we look at the results of the other three techniques for cross-validation of our conclusions. Secondly, we compare the predicitive power of our genetically created model to the best models the other techniques obtained. In this analysis other models are competitors of the genetic model.

5.1 Interpretation of the models

When evaluating the results it is important to keep in mind that the company providing the data is not only interested in good predictive models, but also in conclusions about the most influential variables. Namely, these variables belong to customer features that have the most impact on customer behaviour. Using genetic programming information on the important variables can be obtained by looking at the model with the best fitness. Those variables this model contains 'survived' evolution, and therefore can be assumed to be influential. Nevertheless, there can be differences between the surviving variables and considering only their presence gives no indication of their relative importance. To this end we performed a sensitivity analysis on the variables by fixing all but one variables in the best model and running the value of the free variable through its domain. The changes in the performance of the model are big for very influential variables, and small for variables with less impact. The results of the sensitivity analysis are given in Table 3, where a high value indicates an variable with high influence, while lower values show a lower impact.

duration of relation	4.32%
duration of investment	14.37%
s-ver	13.50%
nr of funds	2.88%
amount invested	18.56%
risk of portfolio	82.35%

Table 3. Results of the sensitivity analysis

Comparing the interpretations of the different techniques, i.e. (dis)agreement on the importance of variables we observed a high level of agreement. Based on this observation it was possible to bring out a well-founded advice for the company that specific risk values in the portfolio substantially raises the chance of ending the business relationship. This conclusion allows the company to adapt its policy and perform directed actions in the form of advising customers to change the risk value of their portfolio.

5.2 Comparison with other techniques

In this section we show a comparison of the best models created by the four techniques. The comparison is based on cumulating their accuracy values over percentiles of the validation data set. Technically, all models rank the data set in such a way that the cases with a higher rank are assumed to be stoppers with a higher probabilty. Thus, the percentage of stoppers is supposed to be the highest in the first percentile. In case of perfect ranking this percentage would be 100. Cumulating over the percentiles means that the 20th cumulative value (horizontal axis in Table 1) equals the percentage of stoppers in the percentiles 1 to 20. Obviously the 100th value equals 38.6 % for each model, since it is the percentage of stoppers in the whole validation data set.

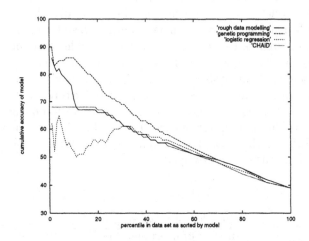

Fig. 1. Predictive power of the best models created by the four techniques

Table 1 shows that the model created by genetic programming highly outperformes models of other techniques. The difference is the biggest in the beginning, and gradually diminishes when cumulating over a high number of percentiles. To this end, it is important to note that the most important percentiles are the first ones that contain the risky customers.

6 Concluding Remarks

In this paper we described an application orieneted research project on applying different modelling techniques in the field of marketing. Our conclusions and recommendations can be summarized as follows.

- It is highly recommendable to use separate training, test and validation data sets. When developing our models we ruled out several models that performed good on the training set, but broke down on the test set. Cross-checking afterwards on the validation set (playing the role of the real-world application data) showed bad performance of these models, confirming the importance of this cautiousness.
- Cross-validation on the most influential variables based on models developed with other techniques raises the level of confidence. In our project we observed good agreement between conclusions of the four research groups.

- Non-linear techniques such as genetic programming and rough data modelling proved to perform better than linear ones with respect to the predictive power of their models on this problem. This observation is in full agreement with the outcomes of an earlier comparative research on a different problem, cf. [1].
- It is advisable to use CoC as fitness measure in the GP. Control runs with accuracy as fitness, as mentioned in section 4, led to models that had a worse accuracy than those evolved with CoC.
- Simple statistical data analysis can distinguish powerful variables. Using this information during the initialization of the GP works as an accelerator, by creating a relatively good initial population.
- It is somewhat surprising that even a long run of the GP could only raise the performance of the initial population by approximately 2 % in terms of accuracy on the training set. Notice, however, that in financial applications one percent gain in predictive performance can mean millions of guilders in cost reduction or profit increase.

The global evaluation of the whole project is very positive. From the perspective of the evolutionary working group the results are good, genetic programming delivered the model with the highest predictive performance. The superiority of the genetically created model is substantial with respect to other models. Furthermore, the gain in the 'density' of stoppers is high, while the whole validation data set contains roughly 39% stoppers, after sorting the data by the genetic model this is 90% in the first percentile and remains above 80% up to the 17th percentile.

The project is also successful from the practical point of view. The company in question agreed that the scientific foundation of the conclusions on behavioural patterns of customers is solid and confirmed the plausibility of the advices. The company was also satisfied with the high accuracy of predicting which customers are possible stoppers. Currently the implementation of an evolutionary data analysis system and its integration into the existing infrastructure is being prepared.

Ongoing and future research concerns adding new types of variables, validation of the models for other time periods as well as developing models for a longer planning horizon. In the meanwhile we are extending the collection of benchmark techniques by including neural networks.

References

1. A.E. Eiben, T.J. Euverman, W. Kowalczyk, E. Peelen, F. Slisser and J.A.M. Wesseling. Comparing Adaptive and Traditional Techniques for Direct Marketing, in H.-J. Zimmermann (ed.), Proceedings of the 4th European Congress on Intelligent Techniques and Soft Computing, Verlag Mainz, Aachen, pp. 434-437, 1996.
2. D. Haughton and S. Oulida, Direct marketing modeling with CART and CHAID, in Journal of direct marketing, volume 7, number 3, 1993.
3. D.W. Hosmer and L. Lemeshow. Applied logistic regression, New York, Wiley, 1989.

4. J. Koza, Genetic Programming, MIT Press, 1992.
5. W. Kowalczyk, Analyzing temporal patterns with rough sets, in H.-J. Zimmermann (ed.), Proceedings of the 4th European Congress on Intelligent Technologies and Soft Computing, Verlag der Augustinus Buchhandlung, pp. 139-143, 1996.
6. M. Magidson, Improved statistical techniques for response modeling, in Journal of direct marketing, volume 2, number 4, 1988.
7. F.F. Reichheld, Learning from Customer Defections, in Harvard Business Review, march-april 1996.
8. R. Walker, D. Barrow, M. Gerrets and E. Haasdijk, Genetic Algorithms in Business, in J. Stender, E. Hillebrand and J. Kingdon (eds.), Genetic Algorithms in Optimisation, Simulation and Modelling, IOS Press, 1994.

An Evolutionary Hybrid Metaheuristic for Solving the Vehicle Routing Problem with Heterogeneous Fleet

Luiz S. Ochi, Dalessandro S. Vianna,
Lúcia M. A. Drummond and André O. Victor

Pós-Grad. Ciência da Computação, UFF
R. São Paulo, 24210-130, Niterói, Rio de Janeiro-Brazil
e-mail:{satoru,dada,lucia,armis}@pgcc.uff.br

Abstract. Nowadays genetic algorithms stand as a trend to solve NP-complete and NP-hard problems. In this paper, we present a new hybrid metaheuristic which combines Genetic Algorithms and Scatter Search coupled with a decomposition-into-petals procedure for solving a class of Vehicle Routing and Scheduling Problems. Its performance is evaluated for a heterogeneous fleet model, which is considered a problem much harder to solve than the homogeneous vehicle routing problem.

1 Introduction

Metaheuristics have proven to be very effective approaches to solving various hard problems [8] [15]. Among them, Genetic Algorithms, Scatter Search and Tabu Search have been used successfully in optimization problems.

Genetic algorithms (GA) were introduced by John Holland and researchers from University of Michigan in 1976 [12]. In the last decade they have become widely used, however, they have not presented very good results in several optimization problems in its original model.

In order to turn GA more efficient, some proposals have appeared recently, such as the inclusion of concepts of scatter search, tabu search and local search in GA ([8] [9] [16]). Particularly, the version of GA with scatter search aims to generate GA "less probabilistic" than conventional ones [8] [9] [16].

In this paper, we propose a new hybrid metaheuristic based on genetic algorithms, scatter search and a decomposition-into-petal procedure to solve a class of vehicle routing problems.

Vehicle routing problems are generalizations of the classical Traveling Salesman Problem (TSP) and consequently are NP-complete problems. Related literature is vast and it usually presents approximate solutions obtained by heuristics. This present popularity can be attributed to the success many algorithms have achieved for a large variety of vehicle routing and scheduling problems in which real systems constraints were implemented. Among them, we include the following papers: [1]; [2]; [3]; [4]; [8]; [9]; [10]; [13]; [14];[16].

In order to evaluate the proposed algorithm, we apply it to a model with a heterogeneous fleet. It can, however, be adjusted for an additional set of constraints which usually arise in real systems. Partial computational results show the efficiency of applying these hybrid techniques in comparison to the algorithm proposed by Taillard [18].

The remainder of this paper is organized as follows. Section 2 describes the Vehicle Routing Problem. Section 3 presents the algorithm proposed. Experimental results are shown in Section 4. Finally, Section 5 is the conclusion.

2 The Vehicle Routing Problem

The Vehicle Routing Problem (VRP) was originally posed by Dantzig and Ramser [5] and may be defined as follows: vehicles with a fixed capacity Q must deliver order quantities q_i $(i = 1, \ldots, n)$ of goods to n customers from a single depot $(i = 0)$. Knowing the distance d_{ij} between customers i and j $(i, j = 0, \ldots, n)$, the objective of the problem is to minimize the total distance traveled by the vehicles in such a way that only one vehicle handles the deliveries for a given customer and the total quantity of goods that a single vehicle delivers is not larger than Q. Several variants of VRPS exist. In the heterogeneous problems, we have a set $\Psi = \{1, \ldots, k\}$ of different vehicle types. A vehicle of type $k \in \Psi$ has a carrying capacity Q_k. The number of vehicles of type k available is n_k. The cost of the travel from customer i to j $(i, j = 0, \ldots, n)$ with a vehicle type k is d_{ijk}. The use of one vehicle of type k implies a fixed cost f_k and different vehicle types will reflect different fixed costs.

In this paper, we are interested in a special case of the above problem where the travel costs are the same for all vehicle types $(d_{ijk} = d_{ijk^1}, k, k^1 \in \Psi)$ and the number n_k of vehicles of each type is not limited $(n_k = \infty, k \in \Psi)$. The goal of this problem is to determine a fleet of vehicles such that the sum of fixed costs and travel costs is minimized.

Although solution methods for homogeneous vehicle routing problems have substantially improved, the vehicle routing problem with heterogeneous fleet has attracted much less attention. This is mainly due to the increased difficulty in solving such a problem [18].

3 An Evolutionary Hybrid Metaheuristic using GA and Scatter Search with Petal Decomposition

The algorithm proposed PeGA (Petal Genetic Algorithm) is a hybrid procedure which uses genetic algorithms, scatter search and a petal decomposition criterion to build chromosomes. This petal decomposition procedure can be used in several models of routing and scheduling problems, such as: the classical model of vehicle routing, routing with heterogeneous fleet and routing with time constraints.

The algorithm proposed executes the following steps, described in next subsections:

- Building of routes/schedules using a petal decomposition procedure

- Representation of solutions by chromosomes
- Generation of an initial population of chromosomes
- Reproduction of chromosomes and their evaluation
- Diversification technique

3.1 Tools for Building of Routes

It has shown in many applications of the vehicle routing problem and its generalizations, that the set of routes has a petal structure which forms a flower. The problem of establishing the best flower is the focus of several papers [6] [17]. We adopted a node "sweeping" technique described as follows.

We associate the VRP to a complete symmetric planar graph $G(N, A)$ where N is the node set and A is the edge set. A cost c_{ij} is associated with each edge, the depot is located at node S and there are n nodes, each associated with one customer. Geometrically, the petals on the plane can be obtained by tracing a vector $\vec{sx}$ with origin in the node S and parallel to axis x. We rotate $\vec{sx}$ in clockwise direction (or counter clockwise direction) until $\vec{sx}$ intercepts a node n_i with demand q_i. Then, we update the accumulated demand $D := q_i$ and go on the rotation until another node is met. For each node met, its demand is added to D and D is compared to the capacity of the vehicle (Q). If it is less than Q, the rotation continues. If D exceeds Q, the last node intercepted by $\vec{sx}$ is deleted from the current petal (route) and the petal is closed. The closing of one petal initiates the process of constructing a new petal, beginning with the deleted node. This procedure is repeated until all nodes are incorporated into petals.

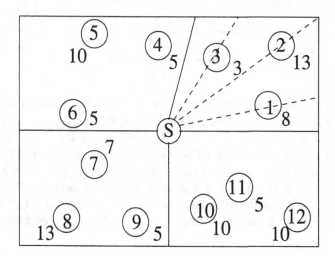

Fig. 1. Building of petals

Figure 1 shows nodes representing customers and their demands, where the capacity of each vehicle is 25. The first petal covers nodes $\{1, 2, 3\}$ whose demands are 8, 13 and 3 respectively. Using the procedure described above, with n nodes, there are $2n$ feasible solutions in petal structures. Of these solutions, n of them are generated by each node in clockwise direction and the remainder is generated in counter clockwise direction. Unfortunately, some solutions of VRP do not have a petal structure. For example, from the original sequence of customers $P_1 = (1\ 2\ 3\ 4\ 5\ 6\ 7\ 8\ 9\ 10\ 11\ 12)$ if we swap 1 and 2, 3 and 6 , and 8 and 12 we obtain a new sequence $P_2 = (2\ 1\ 6\ 4\ 5\ 3\ 7\ 12\ 9\ 10\ 11\ 8)$. If we initiate the petal construction at node 2 in counter clockwise direction, we obtain the following set of petals $\{\{2, 1\}, \{6, 4, 5, 3\}, \{7, 12, 9\}, \{10, 11\}, \{8\}\}$, as shown in Figure 2. This structure is called artificial flower and it is composed of artificial petals.

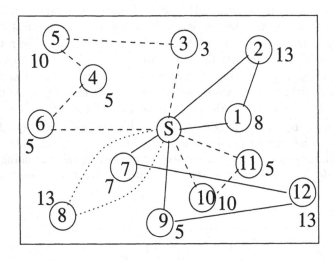

Fig. 2. Artificial petals

We propose the following strategy for the Vehicle Routing Problem with Heterogeneous Fleet. Consider k types of vehicle: $\Psi = \{1, \ldots, k\}$ such that $Q_1 \leq Q_2 \leq \ldots \leq Q_k$ and $f_1 \leq f_2 \leq \ldots \leq f_k$, where Q_i is the capacity of vehicle of type t_i and f_i is its corresponding fixed cost. Each petal is built as follows. The algorithm analyzes the possibilities given by the n types of vehicles and it chooses the type of vehicle t_i which presents the lowest value for $(Q_i - D_i) \times f_i$, where D_i is the accumulated demand of the petal using vehicle t_i. Clearly, the size of each petal will vary according to the type of vehicle associated.

3.2 Coding and Decoding of Chromosomes

In the algorithm PeGA applied to a vehicle routing problem with heterogeneous fleet, a chromosome is a list of k genes, where k is the number of customers.

The ith position (gene) is occupied by a positive integer which represents the customer.

3.3 Population Generation

The size of the initial population in PeGA is a number x such that $n \leq x \leq 2n$, where n is the number of nodes of the associated graph G and chromosomes are generated using the procedure described in Subsection 3.1 for each node in N.

3.4 Genetic Operator and Chromosome Evaluation

The genetic operator is responsible for the production of new chromosomes from the existing ones. The roulette criterion is applied to choose the parents based on the probability associated with each individual according to its fitness. The operator used is an adaptation of the ERX method [19]. For each customer a list of neighbors is built based on its location in the petals of the parents used for reproduction, where the number of parents is greater or equal to 2. For example, in case of two parents and their corresponding set of petals $P_1 = \{\{0,1,3,5,7,2\},\{0,6,4,8,9\}\}$ and $P_2 = \{\{0,3,6,4\},\{0,5,2,1\},\{0,9,7,8\}\}$, the following lists would be obtained for each customer:

- customer (1): 3,2
- customer (2): 7,5,1
- customer (3): 1,5,6
- customer (4): 6,8
- customer (5): 3,7,2
- customer (6): 4,3
- customer (7): 5,2,8,9
- customer (8): 4,9,7
- customer (9): 8,7

In order to generate the offspring the following steps are executed:

Step 1 : the closest customer of the depot v is selected as the first gene of the offspring.

Step 2: a customer w, neighbor of v, with the lowest number of neighbors, is inserted in the chromosome, since it has not yet been selected.

Step 3: $v := w$

Step 4: go to Step 2 until all customers are inserted in the chromosome.

If there is not a customer which is neighboring the last customer inserted (v) and that has not yet been inserted in the chromosome, then the closest customer of v (not inserted yet) is selected.

We use the same criterion described in Subsection 3.1 to select the vehicles which will serve the routes in the offspring, which may be composed by petals or artificial petals.

The chromosome evaluation is based on the cost of the set of associated routes, where each route is considered a TSP and is solved through heuristic GENIUS [7].

3.5 Diversification Technique

The diversification procedure is triggered when the renewal tax (tax of replacement of parents by offspring) is less than 5 per cent. When this occurs, a new population is generated based on the best chromosome. Thus, a "window" is opened in this best chromosome in a random position and pairs of genes are switched in this window. For example, consider the chromosome $\{1,3,5,7,2,6,4,8,9\}$ and the windows:

- $\{\mathbf{1,3,5},7,2,6,4,8,9\}$
- $\{1,3,\mathbf{5,7,2},6,4,8,9\}$
- $\{1,3,5,7,2,6,\mathbf{4,8,9}\}$

After swapping some genes in these windows, we could obtain the chromosomes:

- $\{\mathbf{3,1,5},7,2,6,4,8,9\}$
- $\{1,3,\mathbf{2,7,5},6,4,8,9\}$
- $\{1,3,5,7,2,6,\mathbf{4,9,8}\}$

The vehicles which will serve the routes of these new chromosomes will be chosen using the criterion described in Subsection 3.1.

The algorithm finishes after m diversifications, where m is entry parameter.

4 Experimental Results

Our experiments were run on an IBM RISC/6000 using the programming language C.

Although the Vehicle Routing Problem with Heterogeneous Fleet is a major optimization problem, it has not attracted much attention due to the fact that it is very difficult to solve [18]. Thus in order to analyze PeGA, we used the results presented by Taillard.

Our algorithm was analyzed with the values presented in the table of Figure 3. This table provides information about the problem number (as in Golden *et al* [11]), number of customers (n), vehicle capacities (Q) and their fixed costs (f). An average of results is shown in Figure 4. It presents eight cases where for each of them 10 instances were run. The parameters used were the following:

- number of parents for reproduction = 4
- window size = number of customer / 10
- number of pairs of genes swapped in the window = number of customers / 25
- population size = 2 × number of customers
- number of diversifications = 10

Figure 4 shows that PeGA achieved the best results for the problems 13 and 14, nearly identical results to the Taillard in problems 15, 16, 17 and 18, and presented worst results for the problems 19, 20.

Problem number	n	Vehicle type										
		A		B		C		D		E		F	
		q_A	f_A	q_B	f_B	q_C	f_C	q_D	f_D	q_E	f_E	q_F	f_F
13	50	20	20	30	35	40	50	70	120	120	225	200	400
14	50	100	120	160	1500	300	3500						
15	50	50	100	100	250	160	450						
16	50	40	100	80	200	140	400						
17	75	50	25	120	80	200	150	350	320				
18	75	20	10	50	35	100	100	150	180	250	400	400	800
19	100	100	500	200	1200	300	2100						
20	100	60	100	140	300	200	500						

Fig. 3. Data for VRP with Heterogeneous Fleet

5 Concluding Remarks

Our results so far show some advantages for PeGA when compared to the algorithm presented by Taillard, since GA admit some flexibility considering their parameters adjustment. It means that if more computational time is spent through the increasing of the value of some parameters, such as the population size and the number of generations, the search process is extended, which may improve the quality of the final solution. On the other hand, in some heuristic methods, for example the Tabu Search, the solution met is already the final solution.

Experimental tests will continue and deeper analysis will be conducted for the optimization of the parameters presented in the section above.

We did not use refinements techniques as 2-optimal, which could improve the quality of solution.

As genetic algorithms require a large amount of time to find solution, we are also working on a parallel version of the genetic algorithm for this problem.

Acknowledgments
This work was partially supported by the Brazilian agency CNPq.

References

1. BALL, M.O.; MAGNANTI, T.L.; MONNA, C.L. and NENHAUSER, G.L., 1995, Network Routing, *Handbook in Op. Res. and Manag. Science*, Vol 8, Elsevier Science.
2. BODIN, L.D. and GOLDEN, L., 1981, Classification in Vehicle Routing and Scheduling, *NETWORKS*, Vol 11, 97-108.

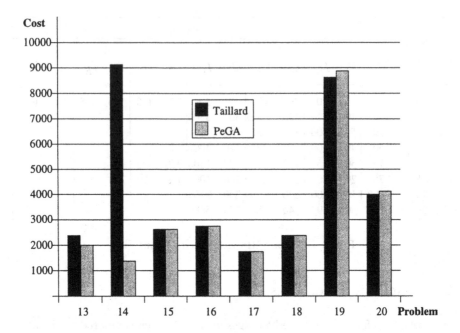

Fig. 4. Experimental Results

3. BODIN, L.D., 1990, Twenty Years of Routing and Scheduling, *Oper. Research*, Vol 38(4), 571-579.
4. BODIN, L.D.; GOLDEN, L.; ASSAD, A.A. and BALL, M., 1993, Routing and Scheduling of Vehicles and Crews, The State of the Art,*Comp. and Op. Res.*, Vol 10, 63-211.
5. DANTZIG, G.B. and RANSER, J.H., 1959, The truck dispatching problem, *Management Science*, Vol 6, 80-91.
6. FOSTER, B.A. and RYAN, D.M., 1976, An Integer Programming Approach to the Vehicle Scheduling Problem, *Op. Res. Quarterly.* 27, 367-384.
7. GENDRAU, M.; HERTZ, A. and LAPORTE, G., 1992, New Insertion Post-Optimization Procedures for the Traveling Salesman Problem, *Oper. Res.* 40, 1086-1094.
8. GLOVER, F., 1995, Scatter Search and Star-Paths: Beyond the Genetic Metaphor, *OR SPECTRUM*, Vol 17, Issue 2/3.
9. GLOVER, F., 1997, Tabu Search and Adaptive Memory Programming: Advances, Applications and Challenges. *In Interfaces in Computer Science and Operations Research*, 1-76, Kluwer Academic Publishers.
10. GOLDEN, B.L., 1993, Special Issue on Vehicle Routing 2000: Advances in Time-Windows, *Optimality, Fast Bounds and Mult_Depot Routing, Am. J. Mgmnt. Sci.*, Vol 13.
11. GOLDEN, B.; ASSAD, A.; LEVY, L. and GHEYSENS, F.G., 1984, The Fleet Size and Mix Vehicle Routing Problem, *Computers and Operations Research*, Vol 11, 49-66.
12. HOLLAND, J.H., 1976, Adaptation in Natural and Artificial Systems, *University of Michigan Press, Ann Arbor.*

13. OCHI, L.S.; DRUMMOND, L.M.A. and FIGUEIREDO, R.M., 1997, Design and Implementation of a Parallel Genetic Algorithm for the Traveling Purchaser Problem, *ACM Symposium on Applied Computing*, San Jose - California, 257-263.
14. OCHI, L.S.; RABELO, P.G. and MACULAN, N., 1997, A New Genetic Metaheuristic for the Clustered Traveling Salesman Problem, *In Proceedings of the II Metaheuristic International Conference (II MIC'97)*, Sophia-France, 59-64.
15. REEVES, R., 1993, Modern Heuristic Technics for Combinatorial Problems, *Blackwell Scientific Publications*.
16. ROCHAT, Y. and TAILLARD, E.D., 1995, Probabilistic Diversification and Intensification on Local Search for Vehicle Routing, *Journal of Heuristics*, 147-167.
17. RYAN, D.M.; HJORRING and GLOVER, F., 1993, Extensions of the Petal Method for Vehicle Routing, *J.Op.Res.*, Vol 44(3), 289-296.
18. TAILLARD, E.D., 1996, A Heuristic Column Generation Method for the Heterogeneous Fleet, *Publication CRT-03-96, Centre de Recherche sur les transports, Université de Montreal.*
19. WHITLEY, D.; STARKWEATHER, T. and SHANER, D., 1991, The traveling Salesman and Sequence Scheduling: Quality Solutions Using Genetic Edge Recombination, *Handbook of Genetic Algorithms*, Van Nostrand Reinhold, NY.

Building a Genetic Programming Framework: The Added-Value of Design Patterns

Tom Lenaerts* and Bernard Manderick**

Adaptive Systems Group, Department of Computer Science,
Vrije Universiteit Brussel, Pleinlaan 2, 1050 Brussels, Belgium

Abstract. A large body of public domain software exists which addresses standard implementations of the Genetic Programming paradigm. Nevertheless researchers are frequently confronted with the lack of flexibility and reusability of the tools when for instance one wants to alter the genotypes representation or the overall behavior of the evolutionary process. This paper addresses the construction of a object-oriented Genetic Programming framework using on design patterns to increase its flexibility and reusability.

1 Introduction

As the field of Genetic Programming (GP) matures, industrial interest increases. For instance, GP has been applied to robot control, telecommunications, pattern recognition and automatic design. The application developers (from now on we will refer to those people as *users*) in those businesses are no GP specialists, they are primarily concerned with the application and not GP itself. Therefore, each user's first instinct will be to reuse existing (public-domain) GP packages. Although those packages solve GP problems efficiently it frequently occurs that the problem at hand needs some special variation of GP which was not provided by the consulted software. Even if the GP package allows special-purpose extensions, it is not always trivial how to extend it and what the repercussions are on the rest of the package. This leaves the user with two options : absorb the entire code and alter (crucial) parts of the system to suit his needs or create his own special-purpose environment to solve the problem.

In this paper we address the design of a reusable object-oriented (OO) framework for GP (GPFrame). Such a framework represents the commonalties in GP and can be tuned to different specialized cases to suit the specific needs of the user. Our goal is to provide a framework which will use a collection of the necessary algorithms and representations presented by the user to solve the problem. Furthermore, the design of the framework needs to be robust to current and maybe future variations of GP. For instance, in the situation where the user changes the representation used by the framework from S-expressions to

---

* Correspondence to tlenaert@vub.ac.be
** Correspondence to bernard@arti.vub.ac.be

graphs or maybe strings as in stack-based GP [Per94]. This change will propagate throughout the entire framework and require other changes in, for instance, the reproduction and initialization operators. In other words, the designer needs to be aware of how the system might change in its lifetime or how it might evolve. If design changes are not taken into account, the designer risks major redesign later on. This redesign will in many cases affect the rest of the package. Moreover crucial changes to the design could easily result in the malfunction of existing applications.

To ensure GPFrame's robustness towards changes we used design patterns. The added-value of design patterns over simple OO reuse (like inheritance and composition) can be compared to the added-value of ADT's, such as stacks and queues, in software design or the added value of block-structured systems over unstructured systems using goto's and gosubs. Design patterns provide a framework designer (from now on we will refer to those people as *designers*)with a mechanism to enforce rules on how to reuse specific parts of a complex object-oriented system. Furthermore, patterns can be used to document the design of the framework. Instead of summarizing all the classes and their relationships you only need to know which design pattern was used to be able to extend a specific part of the framework. This kind of documentation will allow the user to gain faster insight in the workings of the framework and how to reuse it. Experiments in OO research concerning the use of design patterns in frameworks have shown that design patterns allow the designer and user to achieve higher levels of design and code reuse [Joh92,GHJV94,SS95].

In Sect. 2, Sect. 3 and Sect. 4 we will explain why frameworks and design patterns were used to implement GPFrame. Afterwards in Sect. 5 a pattern language will be introduced which outlines in a structured way GPFrame and which design patterns are used. To round up this paper we will report our experiences, future work in Sect. 6 and Sect. 7.

2 Problem Domain

In most papers, GP is described as a variation of genetic algorithms (GA's) [Gol89]. The real difference between both kinds of evolutionary algorithms (EA's) is the genotypes representation, i.e. in GP parse trees or computer programs are evolved [Koz92]. In the beginning, GA's used fixed-length bitstrings to encode solutions for the problem at hand. These encoded solutions were evolved using a generational GA and simple operators like one-point crossover, single-bitmutation and fitness-proportionate selection. Over the years, GA's evolved themselves. Fixed-length bitstrings, changed to variable-length bitstrings, fixed-length real-valued strings, and parse trees. The reproduction operators became more complex, like for instance uniform crossover, or had to be adapted because they became useless on the genotypes representation needed for some experiment. Furthermore the overall evolutionary process could be changed from generational to steady state.

Both EA's also evolved in a direction specific to their needs. For instance, in GP the parse trees tend to grow very large over a number of generations. To limit this growth one can set explicit complexity and depth bounds. Apart from this simple technique of parameter setting, one can incorporate in the GP code techniques like minimum description length [IdGS94]. Or, to increase the performance of GP a specialized technique called stack-based GP was introduced which uses a string based representation and a stack in its evaluation mechanism [Per94].

Over the years, a number of free software systems were made available. Most systems started out as libraries of functions and other components which allowed a researcher to compose his/her experimental software. Following the evolution of both EA's the designer of the system was obliged to alter his software at certain time intervals. This to ensure the compatibility of the system with the requirements of the researcher in the domain. For instance, when automatically defined functions became common ground in GP those software systems had to be adapted to allow this new kind of representation [Koz94]. Such an extension can have major repercussions on the design of the software system. Changes in crucial parts in the software can be made by hacking them into the system resulting in cluttered up code which is difficult to understand and maintain. As a result the system can loose its adaptability, since it can become dedicated to a specific variation. In other words, adding future variations becomes impossible. Alternatively, the designer can rewrite large portions of the software. But then again crucial software changes can result in the malfunction of existing applications. Hence, the designer will have to maintain different versions of his package so that existing applications will not be lost.

Those evolving software packages were implemented in different kinds of programming languages. In the last years interest in OO programming languages, like C++ and Java, has increased. Packages written in those languages are in most cases presented as a library of predefined classes, i.e. toolkits. A *toolkit* is a collection of reusable and related classes designed to provide useful general purpose functionality, e.g. a set of collection classes for stacks, lists, queues and hashtables. Such a toolkit tries to avoid that the user has to re-code common functionalities. The user of the toolkit will have to code up the commonalties in GP or GA incorporating the predefined classes in the toolkit and his own specialized functions and representations. In this process, he has to make sure that the variations he would like to incorporate are compatible with the extensions allowed by the toolkit.

We argued that in essence GP is not that different from GA's. In other words, there is a strong abstract relationship between them. It would therefore be useful to collect these common themes in a system which allows the user to plug in his specialized variations. By doing this, we remove the tedious job from the user to code the commonalties himself. He can concentrate on his specific needs, like specialized functions for fitness evaluation, or reproductional operators, or high-performance representations and present them to the system. In the OO community, such a system is called a framework.

3 Frameworks

A *framework* is defined as a set of co-operating classes that make up a reusable design for a specific software domain [Joh92,GHJV94]. Graphic editors are in many cases used as examples for frameworks since there is a clear view on the common parts of graphic editors and how variations should be added. These editors can be extended to implement specific ones for, e.g., PERT diagrams. It is our opinion that this is the same for EA's like GP. Hence, we would like to design a framework for EA's which can be concretized to a specialized application for GP and later on also for GA's.

Concretely, a framework dictates the architecture of the application, i.e. it defines the overall structure, its partitioning into classes and objects, its key responsibilities and collaborations, and the thread of control. In other words, it filters out what parts are common in the domain and which are problem dependent. A framework can be considered as a puzzle which is almost finished and where you have to put in the remaining pieces to complete the puzzle. But the resulting image can vary depending on which pieces you use.

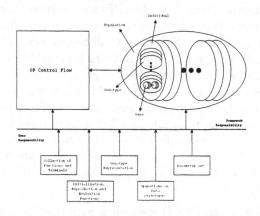

Fig. 1. Conceptual representation of a GP framework.

GPFrame will be responsible for design decisions common to the domain of GP. The user can concentrate on the specifics of the application at hand ; collection of functions and terminals, genotype representation, reproductional operators, overall generational behavior and so on. We consider two parts as common :

1. The *data structures* or aggregate objects representing the populations, individuals, genotypes, genes, functions and terminals. Note though that the kind of terminals and functions can vary depending on the problem. Also a robust framework cannot limit itself to one specific implementation for a genotype.

2. The *GP control flow* which will instantiate the user-defined functions and operators. Thus almost all parts of the GP paradigm should remain variable to allow an user to construct his own GP architecture. The framework will take care of the creation of the functions, their points of entry(where they are called and executed) and how they should be implemented.

Good frameworks are hard to design because the designer gambles that his architecture will work for all applications in the domain. To ensure the robustness of the framework the designer should be familiar with the field and aware of most variations in the domain. Furthermore, the overall structure of the framework should be as flexible as possible to allow current and future variations. Therefore a reuse mechanism is required which allows to increase the flexibility of the framework towards the incorporation of current and future variations.

Reusability in OO languages can be introduced by simple mechanisms like inheritance. Although inheritance allows a certain amount of reuse, it is not sufficient in all cases. For instance, if an application uses a class Set and you would like to change it into an SortedSet. To solve this problem users will make exploit of inheritance. This is a basic OO reuse mechanism which allows him to create a subclass SortedSet which inherits all basic functionalities of the class Set. He will override the methods from Set which should be adapted to suit the behavior of SortedSet, like for instance the method add. This method should now add new elements in a sorted manner. If the user wants to use SortedSet in an application he has to change all instantiations of Set to SortedSet. By hard-coding all instantiations of the class Set or SortedSet in the application, it becomes dedicated on a specific implementation and variations are difficult to introduce . When a complex framework consisting of numerous classes has to be adapted to variations as described above, the maintenance of such a system becomes quite expensive.

A large portion of current OO research is concerned with similar problems, i.e. reuse and flexibility of software packages [Joh92,GHJV94]. One major topic addresses design patterns and pattern languages. Design patterns describe in an abstract and structured way how specific parts of an OO software system can be made flexible and extendible or how to incorporate a certain behavior without dedicating the system to a specific implementation. For instance, in the example of sets and sorted sets, we mentioned that every time you want to use a different set in the application you need to change the name of the class that has to be instantiated. By adding a method called createset to the Set class and its subclasses we can remove all instantiations of the Set or SortedSet from the application. When a new instance of a kind of set is needed, the createSet method will be called. This pattern is called the Factory Method ([GHJV94], p.107). This pattern defines an interface for creating an object and the subclasses decide which class will be instantiated. Through the use of this pattern we can use any kind of set in the application without having to absorb the entire code.

4   Design Patterns

OO design patterns originate from architectural design. C. Alexander introduced a structured collection of patterns, a pattern language, consisting of a set of design patterns which describe in a top-down fashion how worlds, nations, regions, cities, roads, houses, rooms, walls, etc. should be constructed. His pattern language starts from a very high level and works its way down to the finest construction level of walls or windows. The goal of this pattern language was to provide non-architects with a mechanism to design cities or homes.

In this paper we focus on a set of OO design patterns listed by Gamma et al [GHJV94]. The authors address patterns which are descriptions of communicating objects and classes that are customized to solve a general design problem in a particular context. Each pattern represents a common and recurring design solution which can be applied over and over again in different problem-specific contexts.

In general patterns provide the designer with :

1. abstract templates on how to make specific parts of a framework more flexible towards changes.
2. a mechanism to document the architecture of a framework using a high level vocabulary.
3. a mechanism to impose rules on how to reuse or extend the framework, i.e. outline a specific interface on how to incorporate extensions.

and provide the user with :

1. a higher level documentation of a complex framework consisting of numerous heavily interconnected classes and objects
2. a guidance on how to extend the framework with new variations and whether or not the extensions can be made.

Documenting a framework just by summarizing the patterns that were used is not an optimal solution. It would be better to order the patterns in some structured manner as done in Alexander's pattern language for architectural design. Such a pattern language is a kind of roadmap for the user. This roadmap describes how to reuse or extend frameworks with similar design. In our case a pattern language for GPFrame should start out with explaining the use of the framework and identifying the major design topics by pointing to different sub-problems. These sub-problems will then in turn be explained and refer to other sub-sub-problems and so on.

The structure of the pattern language we are aiming at is based on work done by R. Johnson [Joh92]. In his work, he uses a top-down approach to describe a semantic graphic editor called Hotdraw. The entire pattern language is a directed graph with a general pattern as entry-point. The patterns closest to the entry-point are those which are most often used.

Our pattern language which will describe GPFrame, is also a directed graph. In each node, if applicable, we describe the Gamma pattern we used to increase reusability and how it can be reused.

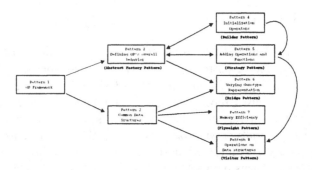

Fig. 2. Graphical representation of the pattern language for GPFrame.

5 A Pattern Language for GPFrame

5.1 Pattern 1 : Genetic Programming Framework

GPFrame is a framework for GP. It can be used to implement specific experiments like symbolic regression, artificial ant problem, induction of decision trees, lawnmower problems and other GP applications [Koz92,Koz94]. The user is able to incorporate simple GP techniques (standard GP) or advanced techniques like automatically defined functions, multiple result producing branches, minimum description length, stack-based GP, demetic grouping and so on [Kin94,Lan96]. The user can extend the application in any way he sees fit.

The two most important parts of the framework are the collection of required data structures and the overall behavior of the GP process. The data structures represent the collection of terminals and functions, the genotypes, the individuals and populations. All operations executed by the evolutionary process are performed on these data structures.

The evolutionary process or the GP control flow is the second important part of the framework. This part provides hooks for problem-specific or specialized functions and operators. The user must provide the system with all operators and functions necessary to perform the evolutionary process. The framework will take care of the functions point's of entry (where they are called and executed on the aggregate objects), and will describe the interface (how certain parts can be extended or reused) of all variable parts. The implementation of for instance crossover or mutation is not hard-coded in a specific class in the framework. The user implements a reproduction operator by creating a new class which satisfies a number of interface prerequisites and connects it to the framework. When the application is executed the framework will instantiate the operator and apply it.

To implement the evolutionary process and its initialization, see **Defining GP's overall behavior (2)**. To ellaborate over the required data structures, see **Common Data Structures (3)**.

5.2   Pattern 2 : Defining GP's Overall Behavior

In GPFrame the overall evolutionary behavior is based on three aspects, a set of GP parameters, a set of GP operators and functions, and information concerning the genotypes representation. There are a number of variations of the latter two aspects. Hence, the user should be able to provide these components without hardwiring them in the framework code.

The GP parameters must be introduced to the framework by a `Parameter` class. This class contains for instance the number of generations and runs, the population size, complexity and depth bounds etc. When an experiment is started, the framework will read an initialization file which consists of all the necessary information and direct this information to the parts of the framework that need it.

Aside from the parameters, the variation in the experiment is determined by the set of GP operators and functions. The set consists of the initialization functions, selection mechanisms, reproduction and post-processing operators, and evaluation functions. To introduce flexibility in GPFrame these operations are not hard-coded into a specific class in the framework but designed as separated classes. For each kind of operator or function their is an abstract class. Concrete subclasses implement different operators, with different behavior. When the user wants to perform an experiment, these GP functions and operators have to be connected to the framework where they can be instantiated and applied. To introduce these operations and functions we made use of the design pattern Abstract Factory (implemented as the `RunFactory` class). *Abstract Factory* provides an interface for creating a set of objects without specifying their concrete classes ([GHJV94], p.87). `RunFactory` contains for all operations a method which creates a new instance of the abstract class type. The framework will call `RunFactory` to obtain a new object of for instance the crossover operator or the fitness evaluation function. Hence, GPFrame remains independent of all problem-specific implementations.

GPFrame allows the user to alter the genotypes representation to a representation required for some specific experiment. To incorporate this kind of flexibility the representation implementation should not be hard-coded in the framework. By removing specific implementation details of the genotype outside the framework, the Abstract Factory pattern is needed again. GPframe uses `GenoTypeFactory` which contains methods to create new instances of the genotype and its problem-specific operations.

To implement the initialization functions, see **Initialization Operators (4)**. To implement the reproduction operators, see **Adding Operations and Functions (5)**. To use different genotype representations, see **Varying Genotype Representation (6)**.

5.3   Pattern 3 : Common Data Structures

In every GP application the containers population, individual, genotype, genes, function set and terminal set are used. A framework definitely needs classes to represent them.

These containers should allow simple GP where each individual consists of one genotype in the population, but also advanced GP where each individual could contain more genotypes. For instance, the individual could consist of one or more result producing branches which can make use of a set of automatically defined functions. GPframe provides the user with 5 main data structures : `Population`, `Individual`, `GenoType`, `MultiTokenSet`(collection of `TokenSet` instances), `TokenSet`(collection of `Tokens`). `TokenSet` consists of `Token` objects which represent problem dependent functions and terminals. Each `Token` object is a collection of terminal and function specific information, i.e. arity, name, type and a reference to a `TokenHandler` object [KM94]. This last object will implement the behavior of a terminal or function for evaluation purposes. Since an individual in an experiment can consist of multiple computer programs it should be possible to define multiple collections of functions and terminals. These collections, i.e. `TokenSet` instances, will be grouped in the `MultiTokenSet` class. Each genotype will be composed of `Gene` objects. These `Gene` instances refer to `Token` objects.

The construction of a new genotype can be done in a naive way, i.e. always create a new `Gene` object and put it in the `GenoType` object. Applying this procedure could have dramatical consequences on the memory efficiency of the framework since the number of `Gene` instances would become very large.

Furthermore, the flexibility of the framework should not be limited to the number of genotypes in an individual. The user should also be able to extend the representation used for the genotype. Ordinary GP makes use of S-expressions which in most cases are represented by parse trees. Alternative representations for these S-expressions like strings or graphs should also be possible.

Application specific algorithms, implemented by the user, will need a number of operations to obtain a result. For instance, crossover could make use of cut and paste operations to execute the process, validator functions to check complexity of the genotype and so on. Again these operations should not be hardwired into the frameworks code since the behavior can vary depending on the problem and the users interest. Hence, the framework needs to incorporate an interface which allows a variety of functions to be performed on the data structures.

To use different genotype representations, see **Varying Genotype Representation (6)**. To elaborate on the memory efficiency of the framework, see **Memory Efficiency (7)**. To incorporate a variety of functions on the data structures, see **Operations on Data Structures (8)**.

5.4   Pattern 4 : Initialization Operators

In GP the populations, individuals and genotypes can be created through different algorithms. For instance, for the construction of these structures Koza initially suggested three techniques ; full-, grow- and "ramped-half-and-half"-initialization [Koz92]. GPFrame allows the user to introduce any of these functions depending on the problem and the representation used for populations, individuals and genotypes. Furthermore, it allows him to introduce a new variation.

This flexibility is incorporated in GPFrame through the use of the Builder Pattern. The *Builder Pattern* allows a user to separate the construction of an aggregate object from its representation ([GHJV94], p.97). This allows the user to use the same construction process to build different representations. For instance, since the genotype's representation can vary, the user needs to implement different constructing mechanism depending on the required representation. Although they vary in the construction process they can all represent the same construction paradigm like for instance grow-initialization. The builder class will implement two methods ; `build` and `get`. The first method will be called to create the specific object and the result will be stored in the class itself. Through the `get` method we obtain the new, fully created object.

GPFrame consists of three builders, `PopulationBuilder`, `Individual-Builder` and `GenoTypeBuilder`. These builders can make use of each other to create an entire population.

Builders in GPFrame also vary in kind. As mentioned earlier different initialization schemes can be incorporated ; full-, grow- and "ramped-half-and-half"-initialization. The user can interchange those different builders depending on the problem or his requirements. Which builder will be used depends on the class that will be instantiated by the `RunFactory`.

Supplying the builders to the framework, see **Defining GP's Overall Behavior (2)**. Introducing different kinds of builders (not related to representation), see **Adding Operations and Functions (5)**.

5.5 Pattern 5 : Adding Operations and Functions

The overall behavior of the evolutionary process is partially determined by the operators and functions used to implement this process. The user should be able to vary these operators and functions depending on his needs.

For instance, the user can provide different crossover operators, selection mechanisms and initialization functions. GPFrame is not aware of the operation it applies, it creates and executes it when needed on the relevant data structures. To obtain this flexibility in function and operator use, GPFrame incorporates the Strategy Pattern. The *Strategy Pattern* allows the user to define a family of related algorithms, encapsulate each one in a different class and use them interchangeably ([GHJV94], p.315).

For instance, to ensure variety in the reproduction operators, GPFrame uses an abstract class `ReproductionFunction`. Through the `RunFactory` the user can supply a set of reproduction operators which are concrete implementations of the abstract class, like for instance the `TreeNodeCrossover` and `SubTreeMutation` class.

Supplying the operations and functions to the framework, see **Defining GP's Overall Behavior (2)**. Using different problem-specific operations in the implementation the operations and functions, see **Operations on Data Structures (8)**.

5.6 Pattern 6 : Varying Genotype Representation

A flexible and reusable framework should allow the user to implement different genotype representations. Thus, GPFrame needs to uncouple the genotype abstraction from its specific implementation. This allows the user to vary the abstraction (result producing branches and automatically defined functions) independently from the implementation (parse trees, graphs or strings as in stack-based GP).

To incorporate this flexibility GPFrame makes use of the Bridge Pattern. The *Bridge Pattern* avoids that there is a permanent binding between the genotype and its implementation leaving both extensible through sub-classing ([GHJV94], p.151). Whenever a new implementation is added it will have no effect on the framework itself. The pattern is implemented through two abstract classes `GenoType` and `GenoTypeImplementor`. The `GenoType` class represents the genotypic concept and contains a number of methods to allow operations on this concept. Each call to `GenoType` is passed on to a specific implementation of the `GenoTypeImplemen-`
tor class. The latter implementation accepts the call, performs some operation and sends back the result (if any).

5.7 Pattern 7 : Memory Efficiency

Inefficient creation of genes to construct for instance a parse-tree can have dramatical consequences on the amount of memory occupied by the entire population. GPFrame should only create a new gene for every function and terminal.

As a solution for this problem GFrame incorporates the Flyweight Pattern. The *Flyweight Pattern* introduces the sharing of `Gene` instances to support a large number of them ([GHJV94], p.195). Through this sharing the total amount of `Gene` objects is reduced significantly. This pattern is implemented by enforcing a one to one relation between `Gene` and `Token` objects (which represent the functions and terminals). In other words, each `Gene` is a unique reference (index) in the `TokenSet` object (Flyweight object).

5.8 Pattern 8 : Operations on Data Structures

When the user creates an algorithm which has to be executed on one of the data structures, he can create other functions on those data structures which can be combined to implement the algorithm. GPFrame cannot anticipate all possible functions from the beginning and therefore needs a mechanism to create any required function.

GPFrame uses the Visitor Pattern to introduce this mechanism. The *Vistor pattern* allows the user to define new operations without changing the classes of elements on which it operates ([GHJV94], p.331). It is particulary well-suited when many unrelated operations shall be implemented on the data structures. The Visitor pattern incorporates this operations in the framework without having to hardwire them in the data structure's class. The class accepts the visitor and calls a specific method in the visitor supplying itself as parameter. This specific method depends on the type of the object. For instance, when the different genotypes in an individual are examined, an instance of `DisplayVisitor` can be sent to each genotype. The genotype accepts the visitor and calls the method `visitGenoType` supplying itself as argument. In the body of that method a display operation is performed, the genotype is displayed on screen and the execution ends.

6 Experience and Future Work

The pattern language presented here is not complete. To use this pattern language the interface on how to implement extensions and a graphical view on the relevant part of the framework should be incorporated. Due to the limited amount of pages we could publish on this subject we could not give an extensive overview on the entire pattern language. Nevertheless, the current result should give the reader an impression of the usefulness of design patterns and a patterns language as reuse mechanism for a generic GP framework. Future work will consist of completing the pattern language to document GPFrame and to extend it to GA's and other EA's.

GPFrame was implemented in Java. This programming language was chosen because of its platform independence and integrated distribution possibilities. The last topic was not addressed in this paper but is currently under investigation. Adding distribution to the frame work should be done in such a way that it does not affect GPFrame. In the OO community this is called separation of concerns, which means that GPFrame should not be concerned with how the distribution is performed [HL95].

To evaluate GPFrame we implemented simple GP problems, like symbolic regression, artificial ant problem and induction of decision trees, and advanced problems, like the two-box problem, the lawnmower problem and generating abstract data types [Koz92,Koz94,Lan96]. In these experiments we used ordinary GP implementations and advanced techniques such as, minimum description length, automatically defined functions, stack-based GP and steady state [Kin94,Koz94]. More experiments have to be done to evaluate further the reusability of the system. One of the current applications is the evaluation of different representations of boolean expressions like reduced ordered binary decision diagrams. Furthermore, to evaluate our pattern language we will present it to students. They will have the opportunity to experiment with the framework and use it for specific applications. Afterwards, they will have to report on the usefulness of the pattern language.

7 Conclusion

GPFrame in its current state was developed over a period of six to nine months. During this period we started out with a GP framework using only simple reuse mechanisms, like inheritance and composition. Using this basic version, we conducted a number of experiments concerning induction of decision trees for data mining purposes. It then became clear that the current GP implementation lacked the flexibility to reuse it for further advanced experiments. Therefore we had to increase its adaptability towards different behaviors and representations. At this point, we incorporated design patterns as a mechanism to facilitate the required flexibility. As a result of this redesign, we noticed that the implementation of new experiments was much easier as before. Furthermore, the framework's architecture was easier to understand and far more robust to changes.

It is our opinion that through the use of these patterns and structuring them into a pattern language, an abstract blueprint can be created on how to design and implement a reusable object-oriented EA's framework.

References

[GHJV94] E. Gamma, R. Helm, R. Johnson, and J. Vlissides. *Design Patterns : elements of reusable object-oriented software*. Addison Wesley, 1994.

[Gol89] D.E. Goldberg. *Genetic Algorithms in Search, Optimization and Machine Learning*. Addison Wesley Publishing Company, 1989.

[HL95] W.L. Hürsch and C.V. Lopes. Separation of concerns. 1995. College of Computer Science, Northeastern University, Boston, USA.

[IdGS94] H. Iba, H. de Garis, and T. Sato. *Advances in Genetic Programming*, chapter 12 Genetic Programming using Minimum Description Length Principle, pages 265 – 284. MIT Press, 1994.

[Joh92] R. Johnson. Documenting frameworks using design patterns. In *Proceedings of OOPSLA '92*. ACM, 1992.

[Kin94] K.E. Kinnear, Jr., editor. *Advances in Genetic Programming*. MIT Press, 1994.

[KM94] M.J. Keith and M.C. Martin. *Advances in Genetic Programming*, chapter 13, Genetic Programming in C++ : Implementation Issues, pages 285–310. MIT Press, 1994.

[Koz92] J.R. Koza. *Genetic Programming : on the programming of computers by means of natural selection*. MIT Press, 1992.

[Koz94] J.R. Koza. *Genetic Programming II : automatic discovery of reusable programs*. MIT Press, 1994.

[Lan96] W.B. Langdon. *Genetic Programming and Data Structures*. PhD thesis, University of London, 1996.

[Per94] T. Perkins. Stack-based genetic programming. In *Proceedings of the 1994 IEEE World Congress on Computational Intelligence*, pages 148–153. IEEE Press, 1994.

[SS95] D. Schmidt and P. Stephenson. Experience using design patterns to evolve communication software across diverse os platforms. In W. Olthoff, editor, *Proceedings Ninth European Conference on Object-Oriented Programming*, pages 399–423. Springer-Verlag, 1995.

Evolutionary Computation and the Tinkerer's Evolving Toolbox

Philip G.K. Reiser
pgr94@aber.ac.uk

Centre for Intelligent Systems
University of Wales
Aberystwyth, Wales, UK

Abstract. In nature, variation mechanisms have evolved that permit increasingly rapid and complex adaptations to the environment. Similarly, it may be observed that evolutionary learning systems are adopting increasingly sophisticated variation mechanisms. In this paper, we draw parallels between the adaptation mechanisms in nature and those in evolutionary learning systems. Extrapolating this trend, we indicate an interesting new direction for future work on evolutionary learning systems.

1 Introduction

It has been argued, [9], that the complexity of life can be accounted for merely by a small number of stochastic processes. They are reproduction, mutation, competition, and selection.

> Neo-Darwinism asserts that the history of the vast majority of life is fully accounted for by only a very few statistical processes acting on and within populations and species [18, p.39]. These processes are reproduction, mutation, competition, and selection. Reproduction is an obvious property of all life. But similarly as obvious, mutation is guaranteed in any system that continuously reproduces itself in a positively entropic universe. Competition and selection become the inescapable consequences of any expanding population constrained to a finite arena. Evolution is then the result of these fundamental interacting stochastic processes as they act on populations, generation after generation [23], [36, p.25] and others. ([9, p.37]).

It is through these elementary processes that the morphology, physiology and behaviour of an organism are adapted to the environment. However, the mechanisms that produce variation have not remained static. Instead, the variation mechanisms have also evolved so that increasingly complex adaptations occur. The resulting trend, the evolution of increasingly sophisticated adaptation mechanisms, has been referred to as the *tinkerer's evolving toolbox*, [27]. In this paper, a selection of variation mechanisms is considered and we investigate how these mechanisms have (or in some cases have not) been modelled in evolutionary learning systems.

The mechanisms considered are (1) mutation (asexual reproduction); (2) genetic recombination (sexual reproduction); (3) nervous system; (4) symbolic reasoning; and (5) language. In actual fact, these variation mechanisms are cumulative, as summarised in Table 1, and so their effect is compounded.

Compound Variation Mechanism
mutation
mutation + recombination
mutation + recombination + nervous system
mutation + recombination + nervous system + reasoning
mutation + recombination + nervous system + reasoning + language

Table 1. The accumulation of variation mechanisms in evolutionary adaptation

In the next sections, we describe each mechanism in turn, and describe evolutionary learning systems that employ these mechanisms.

2 Mutation

The four basic processes identified above are sufficient to allow the genotype to be transformed to cause the organism's phenotype to be highly adapted to its environment. However, the efficiency of adaptation at this basic level is very poor. Truly random mutation is a double-edged sword: as the mutation rate increases, the possible rate of adaptation also increases, but simultaneously, there is a higher risk of destructive mutations occurring.

Consequently, any organism that exhibits a slightly better means of adaptation than a uniform distribution of random mutations will propagate rapidly. Indeed, the processes reported in the biological literature are far more complex than random mutation. For instance, there is increasing evidence to suggest that there exist mechanisms encoded in the genotype that are able regulate the rate of mutation, [27, 31, 30].

Variation through mutation is modelled in Evolutionary Programming, [12, 11], which concerns the simulation of asexual reproduction. Algorithms are based upon the processes of reproduction, selection, competition, and mutation. Mutation is viewed as an operation that causes perturbations while "[maintaining] a behavioural link between each parent and its offspring", [11]. These approaches have been successfully applied to real-world and inherently difficult problems, (e.g. the travelling salesman problem which is NP-hard, [10]).

3 Genetic Recombination

In asexual reproduction, the only changes to the genotype come about by chance mutations—offspring remain genetically similar, if not identical, to the parent. With the evolution of sexual reproduction, offspring genotypes arise not from a number of small changes to the genotype, but are composed of two potentially very different genotypes through genetic recombination. As a result the population diversity influences the degree of change between parent and offspring.

In nature, sexual reproduction suffers disadvantages such as finding a suitable mate, yet it is ubiquitous in higher order organisms. This suggests it is advantageous from the point of view of adaptation.

There are several classes of evolutionary algorithm that are based on genetic recombination. Genetic recombination, or *crossover*, allows good partial solutions, or *building blocks*, to be propagated to subsequent generations. As described by Goldberg's building block hypothesis: "instead of building high-performance strings by trying every conceivable combination, [genetic recombination] constructs better and better strings from the best partial solutions of past samplings", [15, p.41]. We briefly describe several basic models.

The Genetic Algorithm, [21, 15], is distinguished by a fixed-length binary string representation and a crossover operator that does not affect the length of the string. An example is illustrated in Figure 1(a) and some successful applications are described in [16].

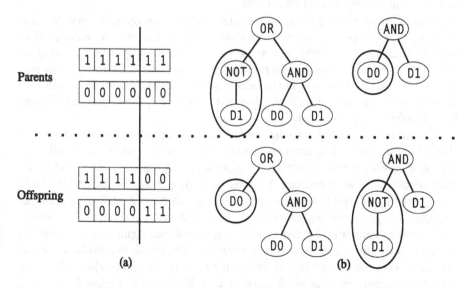

Fig. 1. (a) Crossover of fixed-length binary strings in genetic algorithms; (b) crossover of trees of Lisp S-expressions in genetic programming.

In Genetic Programming, [25], the structures undergoing adaptation are general, hierarchical computer programs of dynamically varying shape and size.

These are expressed as trees of Lisp S-expressions. New programs are primarily produced by moving branches from one tree and inserting them into another. This ensures that programs created by crossover are syntactically valid. Figure 1(b) shows an example of a crossover operation.

Evolutionary Programming, Genetic Algorithms and Genetic Programming are computational models that simulate the processes believed to occur in sexual and asexual reproduction. While many of the details of natural systems have been discarded, the learning algorithms are capable of solving a wide variety of problems, suggesting that some useful processes have been abstracted. However, in the models described so far the organisms, or agents, do not vary their behaviour with respect to their environment. Adaptation only occurs over one or more generations. In the next section, we advance along the evolutionary timescale to consider the mechanisms that enable an organism to adapt its behaviour.

4   Nervous Systems

The information reservoir of an organism comprises not solely of that held in the genotype, but also includes information stored in other ways. For example, information is stored in the firing properties and topology of nerve cells in the nervous system. Changes to nerve cell properties allow the organism's behaviour to modify and adapt during its lifetime.

Genetic adaptation and nervous system adaptation operate over different time scales. Changes to the organism's environment can only be responded to slowly, requiring of the order of generations. The nervous system, on the other hand, is a mechanism that enables rapid adaptive behaviour. Learning in the nervous system allows plasticity in the behaviour of an organism, while genetically-determined behaviour (or instinct) does not vary within the organism's lifetime and therefore tends to be brittle.

Evolutionary learning systems have been constructed that model the evolution of organisms with a nervous system. The nervous system is modelled by a neural network whose structure (or topology) and weights correspond to interconnections between neurons and their firing properties. Neural architectures are evolved by an evolutionary algorithm similar to those described in previous sections. The genotype may encode information about neural network topology, the functions computed by the neuron (e.g. threshold, sigmoid, etc.), and the connection weights, [2]. Over several generations, the population evolves towards genotypes that correspond to high-fitness neural networks. This relation between evolutionary and neural network algorithms is illustrated in Figure 2.

Both genetic algorithms and neural networks have independently demonstrated themselves useful in solving difficult real-world problems. Consequently, the evolution of neural networks has attracted attention in the hope of capturing the benefits of both GA and NNs. The next section, however, will describe a task for which this approach is poorly suited.

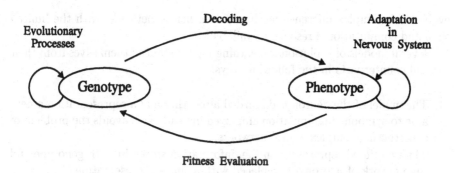

Fig. 2. Evolutionary design of neural networks (adapted from [2]). The genotype is decoded to determine properties of a neural network. The neural network, interacting with the environment, adapts to it; and the better adapted, the greater the probability that the genotype will contribute to the population of the next generation.

5 Symbolic Reasoning

In the 1920s, Wolfgang Köhler studied learning and problem solving in chimpanzees. In a typical experiment, Köhler placed a chimpanzee in an enclosure with a desirable piece of fruit, often a banana, out of reach. The animal had to use nearby objects (e.g. stack several boxes strewn around the enclosure) in order to obtain the fruit. Several interesting observations were made, [1]: (1) rather than being a gradual trial-and-error process, the solution was found suddenly; (2) once solved, repeating the problem resulted in few irrelevant moves; and (3) the animal was able to transfer what it had learned to novel situations.

One popular view is that the animal constructs an internal representation or model of the environment and the objects in it. The model might be a map of the environment or one or more abstract concepts. It is then possible to reason about this model rather than operate on the world itself.

Furthermore, in building a representation of the environment, objects are grouped into sets of properties or concepts. Treating different objects as members of the same concept with the same properties allows different environmental situations to be responded to in a uniform way, thus significantly reducing the complexity of the environment. Concepts may be combined to form propositions, and propositions may be combined to form further propositions. This combining process is commonly referred to as inference.

But how networks of neurons achieve this kind of complex behaviour is still poorly understood and remains a challenging problem for those trying to unravel these phenomena. It is perhaps significant that the evolution of complex reasoning from the nervous system took a very long time even by evolutionary timescales. Unless the process is guided in some way, it is unrealistic to expect to evolve such complex inference mechanisms in neural networks with the limited time and computational resources available.

Several models of evolutionary learning have distanced themselves from their natural counterparts in the following ways.

1. The model of the neuron is discarded altogether and a representation amenable to symbolic manipulation employed instead. This avoids the problem of constructing complex neural networks.
2. The two-tiered representation (i.e. information stored in both genotype and the network of neurons) is replaced with a single representation.[1]

The remainder of this section describes learning systems that make the above two assumptions in order to facilitate reasoning about the environment. Each approach uses a symbolic internal model which is constructed using evolutionary techniques, however, they vary along the following dimensions:

Procedural vs. declarative semantics. Evolving code of conventional programming languages using operators such as crossover and mutation will seldom produce syntactically correct programs, and even less frequently semantically correct ones, [6]. Examples of conventional languages used in evolutionary learning systems include machine language, [13], and state transitions of a finite state machine, [12], as well as high-level languages such as C and Pascal.

Conventional programming languages typically have procedural semantics, and are characterised by a changeable store. The commands of a program influence one another by means of variables held in storage. The relationship between two given commands can be completely understood only with reference to *all* the variables that they both access, and with reference to all the other commands that access these same variables, [33, p. 230]. Clearly, commands are dependent on their context, and operations that change context in an ad hoc manner, such as crossover for instance, will have a poorly understood effect on program meaning.

Representations with declarative semantics on the other hand do not rely on a changeable store and therefore the meaning of a program section is less dependent on its context. This suggests better suitability to manipulation and genetics-based operations. Examples of systems that employ languages with declarative semantics include Learning Classifier Systems, [20, 29], Genetic Programming, [25], and Genetic Logic Programming, [34, 35].

---

[1] A multi-cellular organism stores an instance of the genotype in each cell. Therefore, it is physically very difficult for a multi-cellular creature to modify its genotype rapidly in response to changes in its environment. This difficulty is overcome with behaviour governed by a central organ: the nervous system. Evolutionary change occurs through the culmination of improvements. While this represents a good motivation for having two representations in a natural system, it is not an issue for computer architectures which can flexibly modify data.

Intermediate representation. Approaches may be classified on whether or not they try to fully exploit the implicit parallelism of the genetic algorithm. To draw on the genetic algorithm's implicit parallelism, a representation, typically a subset of first order logic, is associated with a string of elements taken from a binary alphabet. The binary strings are then mapped to a logical expression which is used to reason about the problem domain. For instance,

$$A \ \wedge \ B \ \rightarrow \ C$$

$$\overbrace{\boxed{1\,|\,0}} \quad \overbrace{\boxed{0\,|\,0}} \quad \overbrace{\boxed{1\,|\,1}}$$

Evolving binary strings has the effect of discovering new rules. Examples include Learning Classifier Systems, [20, 29], and genetic concept learning, [7].

The alternative approach is to allow the genetic operators to manipulate a higher level representation directly without an underlying binary representation. Examples include Genetic Programming, [25], and Genetic Logic Programming, [34, 35].

There are arguments both for and against the use of an intermediate representation. As the alphabet size increases, the implicit parallelism (and thus search efficiency) reduces. However, problem domains rarely map naturally onto a binary coding. Furthermore, the choice of encoding is particularly important to ensure good building blocks can be formed. On the other hand, if no intermediate representation is used, then it is easier to handle variable length representations, and the resulting solutions are far more intelligible. However, the price of this is reduced implicit parallelism. The choice of whether to use an intermediate representation is not clear cut.

Operators. The majority of symbolic evolutionary approaches employ operators derived from the genetic algorithm. However, by modifying these operators it is possible to narrow down the search space. For instance, the crossover point may be restricted so that some invalid representations are eliminated.

Other operators are based on the semantics of first-order logic, e.g. generalisation and specialisation operations, that constrain this space in a controlled manner. Examples include [24, 32, 17, 14]. Furthermore, background knowledge supplied by a user can further restrict the search space. For example, operators that perform inductive inference operators to find solutions that are logically consistent and complete with this knowledge, [37, 28].

It is interesting to note that many evolutionary learning systems eliminate the intermediate neural network representation altogether and manipulate symbolic world models directly. The use of first-order logic as the representation language offers several advantages including comprehensible theories, logical inference, and permits a priori (background) knowledge to be exploited.

Several axes have been identified for classifying evolutionary approaches to learning symbolic representations. In the next section language and its role in evolutionary learning is considered.

6 Language

Recall from Section 4 that the genotype does not undergo changes within the lifetime of an organism, and that a secondary representation, embodied in the nervous system, permits more rapid adaptation. Consequently, the information accumulated *during* an organism's lifetime is not genetically propagated to subsequent generations. Instead, this learned behaviour is lost. If, however, an organism were capable of passing on this learned behaviour, and were capable of acquiring experience from another organism, it may well have a selective advantage over other organisms that do not have this capacity. Indeed, mechanisms *have* evolved that allow information acquired during the lifetime of an organism to be propagated.

One such mechanism is the Baldwin effect. In 1896, Baldwin [3] proposed that the ability of an individual to learn could guide the evolutionary process. Furthermore, abilities that initially require learning are eventually replaced by the evolution of genetically determined mechanisms that do not require learning. Consequently, there is a gradual transferral of learned behaviours into the genotype. However, this process is slow, requiring of the order of generations for the learned behaviour to become genetically determined.

Another mechanism that avoids the loss of information acquired during the lifetime of an organism is language. A language provides a common framework for expressing the combination of symbols to describe complex notions such as an organism's environment. It permits the communication of information in a highly structured way and thus allows the exchange of information about symbolic world models. Language therefore provides a mechanism to avoid losing learned information at the end of an organism's lifetime. This description is necessarily oversimplistic; however, our aim is to highlight its role as a mechanism for communicating symbolic world models.

It has been formally shown that communication in multi-agent search can significantly improve search performance over a non-cooperative search, [22, 19]. As learning can be formulated as a search, [26], agent communication may be perceived as a mechanism for accelerating learning. The genetic algorithm has been analysed as a cooperative search process, [4]. The population is viewed as a set of cooperating agents, and a generation is one of many repeated encounters among agents. During crossover, agents communicate parts of their solution (or schemata) to other agents. Test cases were presented that suggest that the genetic algorithm may at times yield a behaviour similar to a cooperative search, [4]. This suggests that crossover can accelerate search because of its role as a mechanism for communication between agents.

There exist learning systems that model agent communication in the form of recombination, as described in Section 3. Yet the literature reports of no evolutionary learning system that employs language or structured communication. A major conclusion therefore drawn is the prediction that introducing structure communication in an evolving population of agents will yield accelerated learning.

7 Conclusion

Natural evolution may be described in terms of the processes of reproduction, mutation, competition, and selection, where mutation refers to the class of mechanisms that introduce variation. The variation mechanism need not be an arbitrary operation, but may perform a complex transformation. Furthermore, there is evidence to suggest that the variation mechanisms themselves are also subject to evolutionary pressure. In natural learning systems, this trend is referred to as the "tinkerer's evolving toolbox".

In this paper, it is proposed that evolutionary learning algorithms may also be characterised by the four processes listed above and therefore share this same framework.

Algorithms are classified according to the complexity of their variation mechanisms. The types of variation, summarised in Table 1, range from chance mutation to complex semantic transformations. It is shown that strong parallels exist between the variation mechanisms of natural evolution and those implemented in evolutionary learning algorithms.

The comparison also reveals that undue significance may be attributed to the mutation and crossover mechanisms as more powerful transformations are possible. Evolutionary learning models that employ a complex representation, such as first order logic, still fit within this framework — and also permit powerful transformations that observe semantic properties.

Finally, it is argued that agent communication serves to accelerate learning. While crossover may be viewed as a form of agent communication, it is unstructured as the information exchanged is chosen arbitrarily. Structured communication between agents therefore represents a promising area worthy of attention.

Acknowledgements

The author would like to thank John Hunt and the anonymous referees for their valuable comments. This work is supported by the University of Wales, Aberystwyth.

References

1. R. L. Atkinson, R.G. Atkinson, E.E. Smith, and D.J. Bem. *Introduction to Psychology*. Harcourt Brace College Publishers, 1993. 11th edition.
2. Karthik Balakrishnan and Vasant Honovar. Evolutionary design of neural architectures – a preliminary guide to the literature. Technical Report CS TR#95-01, Artificial Intelligence Group, Iowa State University, January 1995.
3. J. M. Baldwin. A new factor in evolution. *American Naturalist*, 30:441–451, 1896.
4. Helen G. Cobb. Is the genetic algorithm a cooperative learner? In *Proceedings of the Workshop on the Foundations of Genetic Algorithms and Classifier Systems*, pages 277–296. Morgan Kaufmann, July 1992.

5. W.H.E. Davies and P. Edwards. The communication of inductive inferences. In *Lecture Notes in Artifical Intelligence (1221): Distributed Artificial Intelligence Meets Machine Learning: Learning in Multi-Agent Environments*, pages 223–241. Springer Verlag, Berlin, 1997.

6. Kenneth De Jong. On using genetic algorithms to search program spaces. In John J. Grefenstette, editor, *Genetic Algorithms and their Applications: Proceedings of the second international conference on Genetic Algorithms*, pages 210–216, George Mason University, July 1987. Lawrence Erlbaum Associates.

7. Kenneth A. DeJong, William M. Spears, and Diana F. Gordon. Using genetic algorithms for concept learning. *Machine Learning*, 13:161–188, 1993.

8. Tim Finin, Rich Fritzon, Don McKay, and Robin McEntire. KQML – A language and protocol for knowledge and information exchange. In *Proceedings of the 13th International Workshop on Distributed Artificial Intelligence*, pages 126–136, Seatle, WA, July 1994.

9. David B. Fogel. *Evolutionary Computation: Towards a New Philosophy of Machine Intelligence*. IEEE Press, New York, 1995.

10. D.B. Fogel. Applying evolutionary programming to selected travelling salesman problems. *Cybernetics and Systems*, 63:111–114, 1993.

11. D.B Fogel. Evolutionary programming: an introduction and some current directions. *Statistics and Computing*, 4:113–129, 1994.

12. Lawrence J. Fogel, Alvin J. Owens, and Michael J. Walsh. *Artificial Intelligence Through Simulated Evolution*. John Wiley and Sons, Inc., New York, 1966.

13. R.M. Friedberg. A learning machine: Part I. *IBM Journal of Research*, 2:2–13, 1958.

14. A. Giordana and F. Neri. Search-intensive concept induction. *Evolutionary Computation*, 3(4):375–416, 1995.

15. D. E. Goldberg. *Genetic algorithms in search, optimization, and machine learning*. Addison-Wesley, Reading, MA, 1989.

16. David E. Goldberg. Genetic and evolutionary algorithms come of age. *Communications of the ACM*, Vol. 37:113–119, March 1994.

17. David Perry Greene and Stephen F. Smith. Competition-based induction of decision models from examples. *Machine Learning*, 13:229–257, 1993.

18. A. Hoffman. *Arguments on Evolution: A Paleontologist's Perspective*. Allen and Unwin, London, 1989.

19. Tad Hogg and Bernardo A. Huberman. Better than the rest: The power of cooperation. In L. Nadel and D. Stein, editors, *SFI 1992 Lectures in Complex Systems*, pages 163–184. Addison-Wesley, 1993.

20. J. H. Holland. Escaping brittleness: The possibilities of general-purpose learning algorithms applied to parallel rule-based systems. In T. Mitchell, R. Michalski, and J. Carbonell, editors, *Machine Learning, Volume 2*, chapter 20, pages 593–623. Morgan Kaufmann, San Mateo, CA, 1986.

21. John H. Holland. *Adaptation in Natural and Artificial Systems*. University of Michigan Press, Ann Arbor, 1975.

22. Bernardo A. Huberman. The performance of cooperative processes. *Physica D*, 42:38–47, 1990.

23. J. S. Huxley. The evolutionary process. In J. Huxley, A.C. Hardy, and E.B. Ford, editors, *Evolution as a Process*, pages 9–33. Collier Books, New York, 1963.

24. Cezary Z. Janikow. A knowledge-intensive genetic algorithm for supervised learning. *Machine Learning*, 13:189–228, 1993.

25. John R. Koza. *Genetic Programming: On the Programming of Computers by Natural Selection*. MIT Press, Cambridge, MA, USA, 1992.
26. T.M. Mitchell. Generalization as search. *Artificial Intelligence*, 18(2), March 1982.
27. E. Richard Moxon and David S. Thaler. The tinkerer's evolving toolbox. *Nature*, 387:659–662, 12 June 1997.
28. Philip Reiser. *EVIL1*: a learning system to evolve logical theories. In *Proc. Workshop on Logic Programming and Multi-Agent Systems (International Conference on Logic Programming)*, pages 28–34, July 1997.
29. Stephen F. Smith. *A Learning System Based on Genetic Adaptive Algorithms*. PhD thesis, University of Pittsburgh, 1980.
30. Paul D. Sniegowski, Philip J. Gerrish, and Richard E. Lenski. Evolution of high mutation rates in experimental populations of E. coli. *Nature*, 387:703–705, 12 June 1997.
31. F. Taddei, M. Radman, J. Maynard-Smith, B. Toupance, P.H. Gouyon, and B. Godelle. Role of mutator alleles in adaptive mutation. *Nature*, 387:700–702, 12 June 1997.
32. Gilles Venturini. SIA: a supervised inductive algorithm with genetic search for learning attributes based concepts. In *Proceedings of the European Conference on Machine Learning*, pages 280–296. Springer Verlag, 1993.
33. David A. Watt. *Programming Language Concepts*. Prentice Hall International, Hertfordshire, UK, 1990.
34. Man Leung Wong and Kwong Sak Leung. Inductive logic programming using genetic algorithms. In J.W. Brahan and G.E. Lasker, editors, *Advances in Artificial Intelligence – Theory and Application II*, pages 119–124, 1994.
35. Man Leung Wong and Kwong Sak Leung. The genetic logic programming system. *IEEE Expert Magazine: Intelligent Systems and their Applications*, 10(2):68–76, October 1995.
36. D.E. Wooldridge. *The Mechanical Man: The Physical Basis of Intelligent Life*. McGraw-Hill, New York, 1968.
37. K. Yamamoto, S. Naito, and M. Itoh. Inductive logic programming based on genetic algorithm. *Algorithms, Concurrency and Knowledge*, pages 254–268, 1995.

A Dynamic Lattice to Envolve Hierarchically Shared Subroutines

Alain RACINE[1], Marc SCHOENAUER[1] and Philippe DAGUE[2]

[1] C.M.A.P., Ecole Polytechnique, 91128 PALAISEAU, FRANCE
[2] L.I.P.N., Université PARIS-NORD, 93430 Villetaneuse, FRANCE

Abstract. Our purpose is to enhance performance of Genetic Programming (GP) search. For this, we have been develop a homogeneous system allowing to construct simultaneously a solution and sub-parts of it within a GP framework. This problem is a crucial point in GP research lately since this is intimately linked with building blocks existence problem. Thus, in this paper, we present an "on-going" work concerning $\mathcal{DL}^{GP}$ —Dynamic Lattice Genetic Programming— a new GP system to evolve shared specific modules using a hierarchical cooperative coevolution paradigm. This scheme attempts to improve efficiency of GP by taking one's inspiration of *organization of natural entities*, especially the emergence of complexity. In particular, $\mathcal{DL}^{GP}$ does not require heuristic knowledge. Different credit assignment strategies are presented to compute modules fitness.

$\mathcal{DL}^{GP}$ approach attempts to reduce the global depth of a tree-solution and avoids multiple searches of the same sub-components. Moreover modules induction improves "readability" of GP outputs. In particular, local evolutionary process is applied on the different set of subroutines in order to do converged each population toward a specific ability which remains at disposal of higher level subroutines. Problem decomposition and sub-tasks distribution is emergent through the lattice.

1   Introduction

In his book, *Society of Mind*, Marvin Minsky presents a theory of human thinking and learning. The basis idea is that various phenomena of mind emerge from the interactions among many different kinds of highly evolved brain mechanisms. This conception can be expanded in such a way that *Nature* can be considered as an impressive system where each element –more generally, each concept– is composed by a set of simpler components, this principle been applied recursively. Hence, a natural entity is characterized by a set of sub-components and its own architecture which evolve along three axis : Phylogenetic, Ontogenetic, Epigenetic according to Sipper's paper ([SSM+97]). Moreover, some lower-level elements are shared by multiple encapsulating objects. This can be represented by a lattice corresponding to an abstraction of Nature. Thus, one way to develop a Nature-like system to evolve programs, is to catch this hierarchically homogeneous notion in order to build "smarter" and "versatile" artificial systems.

Moreover, a step toward the reliability of search in genetic programming (GP) is to favor subroutines induction. Indeed evolving primitives set is an efficient way to dynamically reduce complexity. In this way, we should determine off-line a set of low-level functions meeting the condition of *sufficiency*[1]. System dynamically develops new sub-functions allowing to solve more and more complex tasks.

Currently, multiple solutions have been suggested by Koza (*Automatically Defined Functions*), Angeline (*Genetic Library Builder and modules acquisition*) and Rosca (*Adaptive Representation Through Learning*) in particular.

This paper suggests a new GP system inspired from Nature organization, $\mathcal{DL}^{GP}$, that evolves a lattice of high-level primitives from basis components. $\mathcal{DL}^{GP}$ is based on a hierarchical cooperative coevolution paradigm applied to GP. In other words, $\mathcal{DL}^{GP}$ is an evolving subroutines lattice such as each hierarchical layer i describes a set of i-level subroutines populations which evolve locally to hopefully converge toward a specific sub-task solution. This favors *re-usability* of sub-components.

This paradigm follows an homogeneous scheme unlike previous work. Moreover it is not domain-dependant. It does not require more domain specific knowledge than a standard GP. However expert's information can be introduced easily inside the system by adding specific primitives or by tuning parameters within a particular subroutines level.

This paper is a presentation of our work in progress. It is organized as follows. Section 2 overviews several pre-existing approaches promoting subroutines induction. Section 3.1 discusses the necessity of a new paradigm favoring subroutines induction. The framework of our coevolutionary system $\mathcal{DL}^{GP}$ is presented in paragraph 3.2. Different credit-assignment methods are suggested in section 3.4. Benefits and drawbacks of this strategy are discussed in section 4. In last we describe our future work concerning $\mathcal{DL}^{GP}$ experimentation and application to image compression problem.

2 State of the Art

This section presents a survey of several approaches concerning subroutines induction.

2.1 Prehistory

At the beginning of GP age, Koza (in [Koz92]) describes an original approach to optimize programs by means of artificial selection. The evolving individual is a simple tree built up from a static language, typically a Lisp-like language. Koza suggests five major steps to prepare a standard GP run set. The user must determine the *set of terminals*, the *set of primitive functions*, the *fitness measure*, the *parameters for controlling the run* and the *criterion for terminating a run*.

In particular, primitive functions can be either mathematical or boolean expressions, or problem-specific functions.

---

[1] Optimal solution can be expressed by combining of these basis functions.

However, in this framework, a critical growth of the average depth can be observed on account of genetic and selection –probably– operators. Moreover earlier systems didn't use problem decomposability and so didn't take advantage of repeated patterns in the ideal –best– solutions. In such situation, GP spends a lot of time to discover several times same sub-parts for a tree admitting many similar patterns.

To discard this difficulty, we must find a method to decompose the global problem in several relevant subproblems easier to solve. After solving them, the system must recombine sub-solutions to finally obtain the global solution. This methodology could favor multiple uses of the same sub-solution to build the final solution. Thus we hope obtain smaller and more expressive solutions than with standard GP. Several attempts to take into account of problem decomposability will be rapidly surveyed.

2.2 Automatically Defined Functions – ADF

To improve GP performance, Koza ([Koz94]) developed a new paradigm such as *an individual* is both a main program and a set of subroutines which extends the set of initial non-terminal nodes. In fact, subroutines are encapsulated in the same individual as main program.

This strategy requires to specify number and arity of subroutines used by main programs. The best subroutines expansion within the population is realized by means of crossover operator which is modified to respect the new shape of an individual. Crossover takes place at the same context position on both parents (See fig. 1).

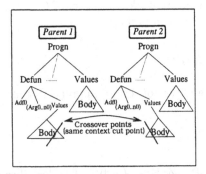

Fig. 1. Context sensitive ADF-crossover

In this model, subroutines are not evaluated against any specific fitness measure but implicitly have the fitness of the main body. Hence, their spreading through generations depends on the fitness of the global program.

The main drawback of ADFs is that a main program which does not perform very well could discard a potentially efficient subroutine : this process does not provide a good measure of the quality of a sub-module. In fact, global fitness of the main program is a too poor information to reflect performance of encapsulated subroutines. Globally, sub-modules evolution process is biased by an irrelevant quantity.

2.3 Genetic Library Builder – GLiB

In order to allow subroutines discovery, Angeline and Pollack ([AP92], [AP93]) suggest a special paradigm to protect best subtrees. For this, they use two additional operators (see fig.2). The first operator, *compression*, produces a subroutine from a random subtree chosen in a global individual. This subroutine is introduced in a particular static (relatively to genetic operator) population called *genetic library* (GLiB). GLiB allows to freeze hopefully useful subtrees and to extend the set of non-terminal primitives. The reverse operator called *expansion* randomly selects a GLiB element and replaces it into the dynamic population.

So, compression operator is a way to propagate a subtree without any distortion through all programs of the population. Expansion operator allows to improve protected sub-modules in replacing them into the active population. In general many individuals use this same subroutine. In this case, it is tested (relatively to selection operator) through a significant quantity of different contexts.

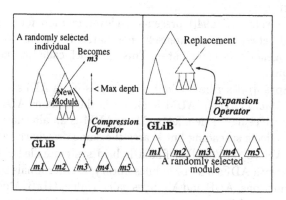

Fig. 2. Compression & Expansion operators

In this approach, the number of chopped off and stored useless subtrees is relatively high. Moreover compression operator seems to lack information to choose the subtree efficiently. So the system might still lose potentially relevant subroutines.

2.4   Adaptative Representations – ARL

Rosca and Ballard ([RB94], ,[Ros95]) describe a method for subroutines induction based on learning. This system called Adaptative Representations through Learning introduces two original operators : subroutine creation and deletion.

Creation operator searches useful building blocks by analyzing the difference between parent and offspring fitnesses and *block activation.* In a way, block activation indicates rate use for a particular block during evaluations on a set of fitness cases. This measure shows block usefulness. From these two informations, creation operator extends vocabulary by generalizing an useful block in a subroutine.

At the opposite, deletion operator discards useless subroutines. Modules utility is computed over a temporal window ; typically the sum of fitnesses of all programs which call it. When a subroutine is deleted, call is replaced by its actual code in any calling program.

This approach based on learning allows to discover useful subroutines by generalizing efficient blocks of code in the population. Nevertheless, modules do not evolve explicitly since there is not local genetic operator —internal to the pool of subroutines— So, building blocks optimization is restricted and useful modules cannot be effectively optimized.

2.5   Other systems

Some other systems have been developed in order to promote useful building blocks emergence.

Especially, Gruau ([Gru94]) presents a clever method for the synthesis of artificial neural networks using intensely modularity by means of developmental process —Ontogenetic development—. This approach called *cellular encoding* is based on grammar.

Spector ([Spe95]) describes a similar system as Koza ADFs called *Automatically Defined Macro* (ADM). ADM behavior is the same as ADF except when functional primitives have side-effects. Thus, ADM is an alternative in this case and avoid side-effects spreading but shares the same drawbacks as ADF.

Teller and Veloso ([TV96b]) define a hybrid system called *PADO*. Given a GP population with ADFs and a Modules library, non-terminals set is composed by primitive functions, ADF and Modules calls. Unlike GLiB, Modules compete explicitly by means of fitness measure based on the weighted average of calling programs. Each generation, the worst library individual is deleted and replaced by the ADFs of the fittest main program.

[TV96a] develops a connectionist representation for evolving programs called *Neural Programming.*

Last, Spector and Stoffel ([SS96]) use specific operators to perform ontogenetic process from a linear coding. These operators organize phenotypical development. *Ontogenetic programming* generates a drastic difference between genotype and phenotype and enhances modularity emergence.

3 $\mathcal{DL}^{GP}$ System Description

3.1 Why a new system ?

Our focus is to enhance GP reliability by reinforcing sub-modules efficiency for a given global task. $\mathcal{DL}^{GP}$ handles completely evolvable modules by using a homogeneous algorithm where each function is undergoing local evolution. In this way, it is not necessary to use specific heuristics requiring domain knowledge. Through this approach, building blocks emergence is favored by protecting useful subtrees from multiple levels.

Indeed, previous approaches lack information to compute subroutines fitness measure and so, general behavior of GP algorithms is biased. More precisely, in these different studies, subroutines don't *evolve* separately of calling programs. So, in the most of these methods, modules evaluate performance from one main body and do not take advantage of multiple contexts of use. Other techniques do not distinguish modules and main body during use of genetic operators. Hence, recombination is anarchic and convergence toward a specific skill is not stable.

We are convinced that a real and emergent distribution of subtasks is more efficient and more homogeneous to solve a global task. Therefore we have developed a distributed hierarchical system with trans-level communication protocol. More precisely, $\mathcal{DL}^{GP}$ optimizes solutions within several layers of subroutines according to local evolutionary processes.

3.2 Informal Description

The approach proposed here is a multi-populations system where a population solves a particular subtask. Sub-problems generation and distribution are emergent. Moreover communication between subroutines is induced implicitly via a fitness measure admitting inter-dependencies. Also, this paradigm corresponds to a coevolutionary system where task decomposition and subtasks solving are performed concurrently.

Hierarchical Topology $\mathcal{DL}^{GP}$ architecture is a hierarchical topology corresponding to a set of strata where a particular layer contains several populations, a population being a pool of subroutines. Each population will hopefully converge toward a specific subtask solution.

Each individual belonging to a specific stratum can call several subroutines located in lower level populations. Thus, subroutines call graph describes a dynamical lattice topology. The user specification of hierarchical distribution constraints the lattice shape. In particular, hierarchical organization avoids recursivity since cyclic calls are forbidden.

So, a lattice node is a particular population and an edge corresponds to a direct call from a subroutine of this population to a lower level one. Hence, a call can jump multiple layers. This means that a subroutine can use any lower level modules. Moreover, lattice edges evolve dynamically and describe a scheme defining current tasks distribution (See fig. 3 presenting an example of lattice).

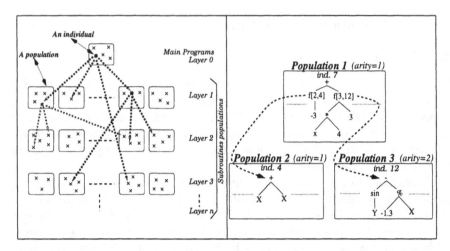

Fig. 3. Lattice Topology & Subroutines calls

Local Evolution To favor the global evolution, each population optimizes its own subtask. So, it seems logical to apply genetic —crossover & mutation— and populational —selection & replacement— operators inside all populations.

Thus each population converges toward a specific ability and provides its skills to higher level ones.

3.3 Genetic operators

Crossover transforms two parents by swapping two randomly selected subtrees taken in each of them. This operator has been defined by Koza in [Koz92]).

Therefore there exists $\mathcal{DL}^{GP}$ trees mutations. Our system allows to define a particular probability of use for each operator such as described by Lamy, Schoenauer and Sebag in [SSJ+96]. This is the list of available mutations :

Random mutation : a subtree is replaced by a new randomly generated one.
Promotion mutation [2] : a node child replaces its parent.
Insertion mutation [2] : a new subtree is inserted, old subtree replaces first child of inserted subtree.
Node mutation 1 : a node is replaced by another admitting a different or not arity.
    Children nodes should be deleted or added according to the arity of the new node.
Node mutation 2 : same as previous but new node arity is the same as old one.
Subroutine mutation : a subroutine call is replaced by another call to a routine of same arity.

The last mutation is $\mathcal{DL}^{GP}$ specific. It allows to try different subroutines calls. In reality, other mutations can produce the same effect but we think that

---

[2] Promotion and Insertion mutations have been introduced by Lamy in [SLJ95].

a particular tuning of this perturbation is necessary in $\mathcal{DL}^{GP}$ to control the coevolutionary process. It allows to search through all lower-level subroutines, the one which best solves same subtask.

3.4 Fitness & Credit Assignment

Like all coevolutionary systems, $\mathcal{DL}^{GP}$ requires to specify a strategy to compute fitness in each population. Indeed $\mathcal{DL}^{GP}$ handles multiple populations whose individual fitnesses interact hierarchically. All populations compute fitness measure from higher level ones. So, we must define a processing order allowing to compute individual performances. $\mathcal{DL}^{GP}$ uses a *top-down* credit-assignment to spread fitness.

First, main population explicitly uses a standard performance computation.

Second level populations compute fitness from higher level fitness values. Each individual knows the list of values transmitted by calling programs and so can process it. Thus each one performs a specific combination (see below for more details) of them. Fitness is the result of this combination.

We can note that tree evaluation is only realized at the top level by main programs. The returned fitness is propagated incrementally through the lattice.

Credit assignment strategies We have seen that a subroutine knows fitnesses list of its calling programs. One can view these credits like rewards or penalties according to overall[3] optimization type : maximization or minimization. So, multiple quantities can be considered to evaluate individual efficiency :

- *Sum* or *mean* of fitnesses of $x\%$ best calling functions
- Fitness of *Best* calling function
- *Variance* of Calling programs fitnesses
- *Number* of calling functions
- *Total number* of calls
- *Rate of calls* per calling program
- Fitnesses beyond a fixed *threshold*

A plenty of credit assignment strategies can be defined by using weighted combinations of these basis measures.

Cooperative Coevolution $\mathcal{DL}^{GP}$ corresponds to a cooperative model. So fitness measure must be "positively depending" between different levels. Basis rule is quite simple :

As a calling program is efficient as its subroutines are strong

---

[3] in keeping with main program optimization

Example This example is based on a main population whose fitness is a cost function (minimization mode). We have defined a fitness criterion for subroutines such as fittest individuals have lowest fitnesses. Thus, $\mathcal{DL}^{GP}$ tries to minimize fitnesses of subroutines.

This is a typical fitness function for subroutines populations.

- Let $\overline{F_{x\%bcf}}$ the means of fitnesses of x% best calling functions
- Let F_{best} the fitness of the best calling function
- Let N_C the number of calls toward this current subroutine
- Let F_{max} the maximal fitness (corresponding to the worst individual)
- Let α, β and γ the weights specified by user
- Let pop such as $1 < pop \leq n$
- Let ind such as $1 \leq ind \leq c_{pop}$
- Let $f[ind, pop]$ is the ind^{th} individual of the pop^{th} population

$$
Fitness(f[ind, pop]) = \begin{cases} \dfrac{1 + \alpha \times \overline{F_{x\%bcf}} + \beta \times F_{best}}{1 + \gamma \times N_C} & \text{if } N_C \neq 0 \\ F_{max} & \text{otherwise} \end{cases}
$$

For main programs (in population 1), fitness evaluation is directly computed from the problem instance.

3.5   Initialization

The first step consists in initializing parameter. In particular, the user prescribes the size of each populations and assigns a specific language. Currently, $\mathcal{DL}^{GP}$ is not typed and it requires compatibles languages. Moreover, homogeneous arity of individuals is inside all populations. Yet, choices between operators from basis vocabulary and subroutines from lower level population is governed by user-supplied rates. Last, a standard maximum depth is specified inside all populations to bound the growth of individuals.

In a second stage, $\mathcal{DL}^{GP}$ initializes randomly all populations from local languages. Process is the same as in standard GP. When a subroutine call is created, a call toward an unknown subroutine but of known arity is built.

3.6   Coevolutionary loop

High-level loop of $\mathcal{DL}^{GP}$ algorithm corresponds to coevolutionary level. It is a top-down evolution of populations in the lattice. In fact, hierarchical dependency induces a particular fitness computing framework. Thus, $\mathcal{DL}^{GP}$ can be divided in two parts : main (or top) population and lower level populations. In this top-down coevolution, each population of the same layer undergoes one or several local evolutionary cycles. Then, the process begins again with populations belonging to the next lower level stratum. This scheme is described in fig. 4. Dashed lines correspond to the local cycles.

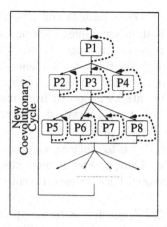

Fig. 4. Coevolutionary cycle

Main population Main population is governed by a standard GP. Genetic operators are ones which are described in section 3.3. Fitness is directly computed from the problem. Of course, the called subroutines are expanded before evaluation of main body. At this level, standard tournament and total replacement are performed.

Other populations For all other populations, evolutionary process is quite different. First, fitness is indirectly computed from fitness of individuals of higher level populations. More precisely, $\mathcal{DL}^{GP}$ evaluates the performance of a subroutine from fitness of programs which call it. For this, $\mathcal{DL}^{GP}$ spreads credits through the lattice. Multiple credit assignments strategies have been viewed in section 3.4.

After fitness of all subroutines has been evaluated, local evolutionary process consists in producing offspring. This stage is crucial and directly influences the efficiency of $\mathcal{DL}^{GP}$. For this, several Selection/Replacement strategies can be defined.

3.7 Selection/Replacement strategy

Two selection/replacement approaches have been defined. Their common basis is that a program calling a particular subroutine, points toward child of this subroutine after any lower level loop.

SSGA-like method $SSGA_{\mathcal{DL}^{GP}}$ selects by tournament one individual in the current subroutines population, applies genetic operators and looks for a site —a parent subroutine— in order to perform replacement. Several strategies are under consideration currently : deterministic replacement by replacing the worst parent and stochastic replacement by using a tournament-like method to choose the best site for the child.

ES-like methods In this scheme, parents and offspring compete for survival. In other words, selection is standard while replacement confronts parents with children. For this, before replacement stage, we must re-compute fitness of offspring. This point is essential. We think a good heuristic consists in comparing parent and its offspring in the same context —calling context—. So, in order to compute offspring credit, fitness is spread from main program but at last, pointer of calling subroutine is turned away from parent toward child. Thus, we can really compare parents and their offspring.

First technique called $ES^{1/1}$ is to replace each parent by one of its direct children (or the parent himself). Thus, pointer is re-directed toward a *true* offspring. Moreover, diversity is promoted because each parent even the worst, produces really one child.

The second method, ES^{mp}, consists in moving pointers of calling functions toward the fittest child of the subroutine parent (See fig. 5). More precisely, all subroutines produce multiple children. Each calling program p which invokes a parent subroutine s, chooses the best offspring of s (in the context of p). In fact, children which survive, are the best in a specific context, ie. in a particular calling program of s. $\mathcal{DL}^{GP}$ uses a garbage collector-like process to discard no more called subroutines (abandoned children and eventually the parent). Moreover, this last scheme requires a population with a flexible size. So, system must use some local constraints in order to bound the growth of the population.

However, the main drawback of these ES-like techniques is the number of fitnesses re-evaluation which explains that is not yet implemented.

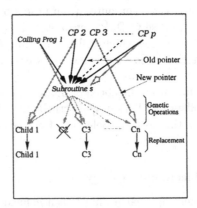

Fig. 5. ES^{mp}: ES with moving pointers and garbage collector process

4   Discussion and further work

$\mathcal{DL}^{GP}$ allows to really evolve subroutines unlike pre-existing paradigms. Each population tries to converge toward a specific ability. Thus $\mathcal{DL}^{GP}$ is a coevolution system and so can be parallelized relatively simply. In fact, local evolutionary process of each population belonging to the same stratum can be computed during the same time. Moreover each population can use a local language to evolve toward a specific ability performing a specific sub-task. So, $\mathcal{DL}^{GP}$ agrees to manage multiples languages. However, this $\mathcal{DL}^{GP}$ version does not accept heterogeneous types. Eventually, this system can solve simultaneously multiples problems sharing several modules. In this case, $\mathcal{DL}^{GP}$ tries to find more relevant modules for all global tasks. Thus "Re-usability" could reach its perfect meaning. For this, it is enough to add other main populations with a specific fitness measure and to propagate fitness through the lattice like we just have seen.

The main drawback of this approach is that fitness evaluation of a subroutine requires a top-down propagation of fitnesses. Future investigations will attempt to decrease temporal complexity, in particular by using a random sampling of main programs selected to spread credit-assignment.

The work presented in this paper is in progress. So, in a first time, further investigations will be focused on $\mathcal{DL}^{GP}$ experimentation in order to evaluate accurately efficiency of its behavior. In particular, we are going to tune credit assignment measure and compare the different selection/replacement heuristics.

Secondly, we will apply $\mathcal{DL}^{GP}$ to the image compression with viewed of testing modularity emergence on problem instances presenting repeated patterns (Yin Yang-like, draught board...). These cases allow to understand visually the "modularity rate" of a specific instance.

Acknowledgment Many thanks to Michèle Sebag and Edmund Ronald for theirs helpful comments about this work.

References

[AP92]   P. J. Angeline and J. B. Pollack. The evolutionary induction of subroutines. In *Proceedings of the Fourteenth Annual Conference of the Cognitive Science Society*, Bloomington, Indiana, USA, 1992. Lawrence Erlbaum.

[AP93]   P. J. Angeline and J. B. Pollack. Coevolving high-level representations. July Technical report 92-PA-COEVOLVE, Laboratory for Artificial Intelligence. The Ohio State University, 1993.

[Gru94] F. Gruau. *Neural Network Synthesis using Cellular Encoding and the Genetic Algorithm*. PhD thesis, Laboratoire de l'Informatique du Parallélisme, Ecole Normale Supérieure de Lyon, France, 1994.

[Koz92] John R. Koza. *Genetic Programming: On the Programming of Computers by Natural Selection*. MIT Press, Cambridge, MA, USA, 1992.

[Koz94] John R. Koza. *Genetic Programming II: Automatic Discovery of Reusable Programs*. MIT Press, Cambridge Massachusetts, May 1994.

[RB94]   J. P. Rosca and D. H. Ballard. Learning by adapting representations in genetic programming. In *Proceedings of the 1994 IEEE World Congress on Computational Intelligence, Orlando, Florida, USA*, Orlando, Florida, USA, 27-29 June 1994. IEEE Press.

[Ros95]   Justinian Rosca. Towards automatic discovery of building blocks in genetic programming. In E. V. Siegel and J. R. Koza, editors, *Working Notes for the AAAI Symposium on Genetic Programming*, pages 78–85, MIT, Cambridge, MA, USA, 10–12 November 1995. AAAI.

[SLJ95]   Marc Schoenauer, Bertrand Lamy, and Francois Jouve. Identification of mechanical behaviour by genetic programming part II: Energy formulation. Technical report, Ecole Polytechnique, 91128 Palaiseau, France, 1995.

[Spe95]   Lee Spector. Evolving control structures with automatically defined macros. In E. V. Siegel and J. R. Koza, editors, *Working Notes for the AAAI Symposium on Genetic Programming*, pages 99–105, MIT, Cambridge, MA, USA, 10–12 November 1995. AAAI.

[SS96]   Lee Spector and Kilian Stoffel. Ontogenetic programming. In John R. Koza, David E. Goldberg, David B. Fogel, and Rick L. Riolo, editors, *Genetic Programming 1996: Proceedings of the First Annual Conference*, pages 394–399, Stanford University, CA, USA, 28–31 July 1996. MIT Press.

[SSJ+96]   Marc Schoenauer, Michele Sebag, Francois Jouve, Bertrand Lamy, and Habibou Maitournam. Evolutionary identification of macro-mechanical models. In Peter J. Angeline and K. E. Kinnear, Jr., editors, *Advances in Genetic Programming 2*, chapter 23, pages 467–488. MIT Press, Cambridge, MA, USA, 1996.

[SSM+97]   Moshe Sipper, Eduardo Sanchez, Daniel Mange, Marco Tomassini, Andres Perez-Uribe, and Andre Stauffer. The POE model of bio-inspired hardware systems: A short introduction. In John R. Koza, Kalyanmoy Deb, Marco Dorigo, David B. Fogel, Max Garzon, Hitoshi Iba, and Rick L. Riolo, editors, *Genetic Programming 1997: Proceedings of the Second Annual Conference*, page 510, Stanford University, CA, USA, 13-16 July 1997. Morgan Kaufmann.

[TV96a]   Astro Teller and Manuela Veloso. Neural programming and an internal reinforcement policy. In John R. Koza, editor, *Late Breaking Papers at the Genetic Programming 1996 Conference Stanford University July 28-31, 1996*, pages 186–192, Stanford University, CA, USA, 28–31 July 1996. Stanford Bookstore.

[TV96b]   Astro Teller and Manuela Veloso. PADO: A new learning architecture for object recognition. In Katsushi Ikeuchi and Manuela Veloso, editors, *Symbolic Visual Learning*, pages 81–116. Oxford University Press, 1996.

Lecture Notes in Computer Science

For information about Vols. 1–1308

please contact your bookseller or Springer-Verlag

Vol. 1344: C. Ausnit-Hood, K.A. Johnson, R.G. Pettit, IV, S.B. Opdahl (Eds.), Ada 95 – Quality and Style. XV, 292 pages. 1997.

Vol. 1345: R.K. Shyamasundar, K. Ueda (Eds.), Advances in Computing Science - ASIAN'97. Proceedings, 1997. XIII, 387 pages. 1997.

Vol. 1346: S. Ramesh, G. Sivakumar (Eds.), Foundations of Software Technology and Theoretical Computer Science. Proceedings, 1997. XI, 343 pages. 1997.

Vol. 1347: E. Ahronovitz, C. Fiorio (Eds.), Discrete Geometry for Computer Imagery. Proceedings, 1997. X, 255 pages. 1997.

Vol. 1348: S. Steel, R. Alami (Eds.), Recent Advances in AI Planning. Proceedings, 1997. IX, 454 pages. 1997. (Subseries LNAI).

Vol. 1349: M. Johnson (Ed.), Algebraic Methodology and Software Technology. Proceedings, 1997. X, 594 pages. 1997.

Vol. 1350: H.W. Leong, H. Imai, S. Jain (Eds.), Algorithms and Computation. Proceedings, 1997. XV, 426 pages. 1997.

Vol. 1351: R. Chin, T.-C. Pong (Eds.), Computer Vision – ACCV'98. Proceedings Vol. I, 1998. XXIV, 761 pages. 1997.

Vol. 1352: R. Chin, T.-C. Pong (Eds.), Computer Vision – ACCV'98. Proceedings Vol. II, 1998. XXIV, 757 pages. 1997.

Vol. 1353: G. BiBattista (Ed.), Graph Drawing. Proceedings, 1997. XII, 448 pages. 1997.

Vol. 1354: O. Burkart, Automatic Verification of Sequential Infinite-State Processes. X, 163 pages. 1997.

Vol. 1355: M. Darnell (Ed.), Cryptography and Coding. Proceedings, 1997. IX, 335 pages. 1997.

Vol. 1356: A. Danthine, Ch. Diot (Eds.), From Multimedia Services to Network Services. Proceedings, 1997. XII, 180 pages. 1997.

Vol. 1357: J. Bosch, S. Mitchell (Eds.), Object-Oriented Technology. Proceedings, 1997. XIV, 555 pages. 1998.

Vol. 1358: B. Thalheim, L. Libkin (Eds.), Semantics in Databases. XI, 265 pages. 1998.

Vol. 1360: D. Wang (Ed.), Automated Deduction in Geometry. Proceedings, 1996. VII, 235 pages. 1998. (Subseries LNAI).

Vol. 1361: B. Christianson, B. Crispo, M. Lomas, M. Roe (Eds.), Security Protocols. Proceedings, 1997. VIII, 217 pages. 1998.

Vol. 1362: D.K. Panda, C.B. Stunkel (Eds.), Network-Based Parallel Computing. Proceedings, 1998. X, 247 pages. 1998.

Vol. 1363: J.-K. Hao, E. Lutton, E. Ronald, M. Schoenauer, D. Snyers (Eds.), Artificial Evolution. XI, 349 pages. 1998.

Vol. 1364: W. Conen, G. Neumann (Eds.), Coordination Technology for Collaborative Applications. VIII, 282 pages. 1998.

Vol. 1365: M.P. Singh, A. Rao, M.J. Wooldridge (Eds.), Intelligent Agents IV. Proceedings, 1997. XII, 351 pages. 1998. (Subseries LNAI).

Vol. 1367: E.W. Mayr, H.J. Prömel, A. Steger (Eds.), Lectures on Proof Verification and Approximation Algorithms. XII, 344 pages. 1998.

Vol. 1368: Y. Masunaga, T. Katayama, M. Tsukamoto (Eds.), Worldwide Computing and Its Applications — WWCA'98. Proceedings, 1998. XIV, 473 pages. 1998.

Vol. 1370: N.A. Streitz, S. Konomi, H.-J. Burkhardt (Eds.), Cooperative Buildings. Proceedings, 1998. XI, 267 pages. 1998.

Vol. 1372: S. Vaudenay (Ed.), Fast Software Encryption. Proceedings, 1998. VIII, 297 pages. 1998.

Vol. 1373: M. Morvan, C. Meinel, D. Krob (Eds.), STACS 98. Proceedings, 1998. XV, 630 pages. 1998.

Vol. 1374: H. Bunt, R.-J. Beun, T. Borghuis (Eds.), Multimodal Human-Computer Communication. VIII, 345 pages. 1998. (Subseries LNAI).

Vol. 1375: R. D. Hersch, J. André, H. Brown (Eds.), Electronic Publishing, Artistic Imaging, and Digital Typography. Proceedings, 1998. XIII, 575 pages. 1998.

Vol. 1376: F. Parisi Presicce (Ed.), Recent Trends in Algebraic Development Techniques. Proceedings, 1997. VIII, 435 pages. 1998.

Vol. 1377: H.-J. Schek, F. Saltor, I. Ramos, G. Alonso (Eds.), Advances in Database Technology – EDBT'98. Proceedings, 1998. XII, 515 pages. 1998.

Vol. 1378: M. Nivat (Ed.), Foundations of Software Science and Computation Structures. Proceedings, 1998. X, 289 pages. 1998.

Vol. 1379: T. Nipkow (Ed.), Rewriting Techniques and Applications. Proceedings, 1998. X, 343 pages. 1998.

Vol. 1380: C.L. Lucchesi, A.V. Moura (Eds.), LATIN'98: Theoretical Informatics. Proceedings, 1998. XI, 391 pages. 1998.

Vol. 1381: C. Hankin (Ed.), Programming Languages and Systems. Proceedings, 1998. X, 283 pages. 1998.

Vol. 1382: E. Astesiano (Ed.), Fundamental Approaches to Software Engineering. Proceedings, 1998. XII, 331 pages. 1998.

Vol. 1383: K. Koskimies (Ed.), Compiler Construction. Proceedings, 1998. X, 309 pages. 1998.

Vol. 1384: B. Steffen (Ed.), Tools and Algorithms for the Construction and Analysis of Systems. Proceedings, 1998. XIII, 457 pages. 1998.

Vol. 1385: T. Margaria, B. Steffen, R. Rückert, J. Posegga (Eds.), Services and Visualization. Proceedings, 1997/1998. XII, 323 pages. 1998.

Vol. 1386: T.A. Henzinger, S. Sastry (Eds.), Hybrid Systems: Computation and Control. Proceedings, 1998. VIII, 417 pages. 1998.

Vol. 1387: C. Lee Giles, M. Gori (Eds.), Adaptive Processing of Sequences and Data Structures. Proceedings, 1997. XII, 434 pages. 1998. (Subseries LNAI).

Vol. 1388: J. Rolim (Ed.), Parallel and Distributed Processing. Proceedings, 1998. XVII, 1168 pages. 1998.

Vol. 1389: K. Tombre, A.K. Chhabra (Eds.), Graphics Recognition. Proceedings, 1997. XII, 421 pages. 1998.

Vol. 1391: W. Banzhaf, R. Poli, M. Schoenauer, T.C. Fogarty (Eds.), Genetic Programming. Proceedings, 1998. X, 232 pages. 1998.